OTHER TELECOMMUNICATIONS BOOKS FROM AUERBACH

Architecting the Telecommunication Evolution: Toward Converged Network Services
Vijay K. Gurbani and Xian-He Sun
ISBN: 0-8493-9567-4

Business Strategies for the Next-Generation Network
Nigel Seel
ISBN: 0-8493-8035-9

Chaos Applications in Telecommunications
Peter Stavroulakis
ISBN: 0-8493-3832-8

Context-Aware Pervasive Systems: Architectures for a New Breed of Applications
Seng Loke
ISBN: 0-8493-7255-0

Fundamentals of DSL Technology
Philip Golden, Herve Dedieu, Krista S Jacobsen
ISBN: 0-8493-1913-7

Introduction to Mobile Communications: Technology,, Services, Markets
Tony Wakefield, Dave McNally, David Bowler, Alan Mayne
ISBN: 1-4200-4653-5

IP Multimedia Subsystem: Service Infrastructure to Converge NGN, 3G and the Internet
Rebecca Copeland
ISBN: 0-8493-9250-0

MPLS for Metropolitan Area Networks
Nam-Kee Tan
ISBN: 0-8493-2212-X

Performance Modeling and Analysis of Bluetooth Networks: Polling, Scheduling, and Traffic Control
Jelena Misic and Vojislav B Misic
ISBN: 0-8493-3157-9

A Practical Guide to Content Delivery Networks
Gilbert Held
ISBN: 0-8493-3649-X

Resource, Mobility, and Security Management in Wireless Networks and Mobile Communications
Yan Zhang, Honglin Hu, and Masayuki Fujise
ISBN: 0-8493-8036-7

Security in Distributed, Grid, Mobile, and Pervasive Computing
Yang Xiao
ISBN: 0-8493-7921-0

TCP Performance over UMTS-HSDPA Systems
Mohamad Assaad and Djamal Zeghlache
ISBN: 0-8493-6838-3

Testing Integrated QoS of VoIP: Packets to Perceptual Voice Quality
Vlatko Lipovac
ISBN: 0-8493-3521-3

The Handbook of Mobile Middleware
Paolo Bellavista and Antonio Corradi
ISBN: 0-8493-3833-6

Traffic Management in IP-Based Communications
Trinh Anh Tuan
ISBN: 0-8493-9577-1

Understanding Broadband over Power Line
Gilbert Held
ISBN: 0-8493-9846-0

Understanding IPTV
Gilbert Held
ISBN: 0-8493-7415-4

WiMAX: A Wireless Technology Revolution
G.S.V. Radha Krishna Rao, G. Radhamani
ISBN: 0-8493-7059-0

WiMAX: Taking Wireless to the MAX
Deepak Pareek
ISBN: 0-8493-7186-4

Wireless Mesh Networking: Architectures, Protocols and Standards
Yan Zhang, Jijun Luo and Honglin Hu
ISBN: 0-8493-7399-9

Wireless Mesh Networks
Gilbert Held
ISBN: 0-8493-2960-4

AUERBACH PUBLICATIONS

www.auerbach-publications.com
To Order Call: 1-800-272-7737 • Fax: 1-800-374-3401
E-mail: orders@crcpress.com

Handbook of
IPv4 to IPv6
Transition

Methodologies for
Institutional and
Corporate Networks

John J. Amoss
Daniel Minoli

Auerbach Publications
Taylor & Francis Group
Boca Raton   New York

Auerbach Publications is an imprint of the
Taylor & Francis Group, an **informa** business

Auerbach Publications
Taylor & Francis Group
6000 Broken Sound Parkway NW, Suite 300
Boca Raton, FL 33487-2742

© 2008 by Taylor & Francis Group, LLC
Auerbach is an imprint of Taylor & Francis Group, an Informa business

Library of Congress Cataloging-in-Publication Data
Amoss, John, 1941-
Handbook of IPv4 to IPv6 transition : methodologies for institutional and corporate networks / John J. Amoss, Dan Minoli.
p. cm.
Includes bibliographical references and index.
ISBN-13: 978-0-8493-8516-2 (alk. paper)
ISBN-10: 0-8493-8516-4 (alk. paper)
1. TCP/IP (Computer network protocol)--Handbooks, manuals, etc. 2. Computer networks--Handbooks, manuals, etc. I. Minoli, Daniel, 1952- II. Title.
TK5105.585.A465 2011
004.6'2--dc22   2007028070

Visit the Taylor & Francis Web site at
http://www.taylorandfrancis.com

and the Auerbach Web site at
http://www.auerbach-publications.com

For Andrew, Olivia, and Amelia — future Internet users.

John

For Anna and the Kids.

Dan

Contents

Foreword

This book requires some upfront knowledge of the Internet Protocol layer and awareness of the concern that version 4 of the Internet Protocol (IPv4) is running out of address space by 2010-2011 with a symbolic date of 10/10/10.

The original work on the Internet design begun in 1973 by Vint Cerf and Bob Kahn benefitted from the collective experience of the predecessor to Internet, the Arpanet. The design of IPv4 took place over the period from 1973 to 1978. It was the product of a recurring series of specifications, implementations and tests that ultimately led to standardization of IPv4 in mid-1978. By the early 1990s it was feared that the rate of consumption of IPv4 address space and the relative inefficiency of its assignment would exhaust the resource within a few years. Work was initiated within the IETF (Internet Engineering Task Force) in 1992 to develop a new version with a larger address space and a feature set that benefited from the many years of experience with IPv4. There ensued a great deal of debate and many different proposals. Ultimately, IPv6 was standardized in 1998.

The IPv6 Forum was created by the members of the IETF IPv6 Working Group (WG) and the Deployment WG led then by Jim Bound who supported my proposal at the IETF IPv6 WG interim meeting on February 5, 1999 in Grenoble and then at the IETF meeting in Minneapolis in April 1999. The proposal was finally adopted and launched in May 1999. The IPv6 Forum is the only body endorsed by the IAB (Internet Architecture Board), the IETF IPv6 WG, and the Internet Society (ISOC) to promote IPv6 worldwide. Dr. Cerf has joined this initiative as its honorary chairman whose mandate is to strengthen its mission.

IPv6 will be largely driven by technology refresh and technology/business case. IPv6 was designed to cater to many deployment scenarios, starting with the extension of the packet technology supporting IPv4 with transition models to keep IPv4 working indefinitely. Scenarios then cater to new uses and new models that require a combination of features that were not tightly designed or scalable in IPv4 such as IP mobility, end to end connectivity, end to end services and ad hoc services, to the extreme scenario where IP becomes a commodity service enabling lowest cost deployment of large scale sensor networks, RFID, IP in the car, to any imaginable scenario where networking adds value to commodity.

IPv6 readiness is a lowest cost option since it is part of a technology refresh and makes the network future-proof (though a careful review of the firewall security is called for). There is also an educational process involved. Again, the scenarios are quite varied and no size fits all. The geopolitical dimension is also crucial for any country to stay or remain among the most advanced IT nations in the world. The simplest scenario is that international companies that deal with Asia, for example, would be required to support the new protocol.

This book takes you through the technology and issues associated with the implementation of IPv6. What is important to recognize is that not all the issues are fully resolved. The authors have succeeded in bringing clarity and scope of knowledge to the reader to enable use of the content in a pragmatic way allowing him to move forward and deploy IPv6 in real life situations.

Latif Ladid
IPv6 Forum President

Preface

Internet Protocol version 6 (IPv6) offers the potential of achieving the scalability, reachability, end-to-end interworking, quality of service (QoS), and commercial-grade robustness for data as well as for Voice-over-IP (VoIP)/triple-play networks. Such capabilities are mandatory mileposts of the technology if it is to replace the time division multiplexing (TDM) infrastructure around the world.

IPv6 is now gaining momentum globally, with major interest and activity in Europe and Asia, and there also is some traction in the United States. For example, the U.S. Department of Defense (DoD) announced in 2003 that from October 1, 2003, all new developments and procurements needed to be IPv6-capable. The DoD's goal is to complete the transition to IPv6 for all intra- and internetworking across the agency by 2008. In 2005, the U.S. Government Accountability Office (GAO) recommended that all agencies become proactive in planning a coherent transition to IPv6. Corporations and institutions need to start planning at this time how to kick off the transition planning process and determine best how coexistence can be maintained during the three- to six-year window that will likely be required to achieve the global worldwide transition. This book addresses the migration and macro-level scalability requirements for this transition.

After an introduction in Chapter 1, in Chapter 2 we provide a brief tutorial of the IPv6 addressing capabilities. Chapter 3 looks at IPv6 network constructs, specifically key routing processes; Chapter 4 examines the IPv6 autoconfiguration techniques. To wrap up this portion of the text, Chapter 5 provides a more formal look at the suite of IPv6-related protocols.

Chapter 6 starts the major discussion theme of this text: IPv6 enterprise/institutional network migration scenarios (tunneling and encapsulation). Coexistence issues are also discussed. Chapter 7 concerns the various elements in the network and what migration role they need to play to support the transition. Chapter 8 looks at actual transition strategies for institutional and enterprise networks. Chapter 9 presents application aspects of the IPv6 transition. Chapter 10 concludes the discussion by looking at security in IPv6 networks.

This book should prove useful to strategic planners at enterprise firms, carriers, and institutions. It will also be useful to software and applications developers.

Authors

John J. Amoss is a distinguished member of the technical staff at Alcatel-Lucent, where he is responsible for wireless data networking strategy. He previously was a member of the Bellcore (now Telcordia) technical staff. Dr. Amoss is one of only seven individuals who received the 1994 Bellcore President's Award for their technical contributions. His diverse responsibilities have included working on a proposed national network for NASA, developing requirements and architecture for business data networks, defining public network data services, addressing potential data service revenues, providing technical descriptions of public network-based Asynchronous Transfer Mode (ATM) services, and developing technical requirements for a messaging platform based on X.400 and X.500 technology. Dr. Amoss has been a participant in the IPv6 Forum, presenting a talk at the Dubai conference. He received the bachelor of science degree (1963), the master of science degree (1968), and the doctor of philosophy degree (1972) from the Johns Hopkins University. Dr. Amoss is an adjunct professor at Stevens Institute of Technology and served as a communications consultant for DataPro (now The Gartner Group). He has coauthored the book, *IP Applications with ATM*, part of the McGraw-Hill Series on Computer Communications. He is also a coauthor of the *CRC Handbook of Modern Telecommunications*. He served as a captain in the U.S. Army from September 1968 through September 1970.

Dan Minoli has many years of telecom, networking, and IT experience for end users, carriers, academia, and venture capitalists, including work at AIG, Prudential Securities, Capital One Financial, Advanced Research Projects Agency (ARPA) think tanks, Bell Telephone Laboratories, ITT, Bell Communications Research (Bellcore/Telcordia), AT&T, Leading Edge Networks Inc., SES Americom, New York University, Rutgers University, Stevens Institute of Technology, and Societé General de Financiament de Québec. Recently, he also played a founding role in the launching of two networking companies through the high-tech incubator Leading Edge Networks Inc., which he ran in the early 2000s: Global Wireless Services, a provider of broadband hotspot mobile Internet and hotspot Wi-Fi VoIP services to high-end marinas; and InfoPort Communications Group, an optical and Gigabit

Ethernet metropolitan carrier supporting data center/storage area network/channel extension and Grid Computing network access services. He is currently working on IPTV systems engineering and deployment, where IP Multicast plays a key role.

An author of a number of technical references and handbooks on information technology, telecommunications, data communications networking, nanotechnology, and sensor applications for Homeland Security, Minoli has also written columns for over 20 years for *ComputerWorld*, *NetworkWorld*, and *Network Computing* (1985–present). He authored the first-ever book on VoIP, *Delivering Voice Over IP Networks* (Wiley, 1998), and first-ever book on VoIP over IPv6, *VoIP Over IPv6* (Elsevier, 2006). Minoli has taught graduate/undergraduate programs at New York University (Information Technology Institute), Rutgers University, and Stevens Institute of Technology (1984–2003). Also, he was a technology analyst at-large for Gartner/DataPro (1985–2001); based on extensive hands-on work at financial firms and carriers, he tracked technologies and wrote around 50 distinct chief technology officer/chief information officer-level technical/architectural scans in the area of telephony and data communications systems, including topics on security, disaster recovery, IT outsourcing, network management, local-area networks (LANs), wide-area networks (ATM and Multi-Protocol Label Switching [MPLS]), wireless (LAN, public hotspot Wi-Fi, and wireless sensor technology), VoIP, network design/economics, carrier networks (such as metro Ethernet and coarse wavelength division multiplexing/dense wavelength division multiplexing [CWDM/DWDM]), and E-Commerce. Over the years, he has advised venture capitalists for investments of $150 million in a dozen high-tech companies and has acted as expert witness in a winning $11 billion lawsuit regarding an early wireless air-to-ground communication system that made use of voice-over-packet technologies in the airplane cabin.

Chapter 1

Introduction and Overview

1.1 Opportunities Offered by IPv6

The Internet Protocol version 6 (IPv6) is now gaining momentum as an improved network layer protocol. There is much commercial interest and activity in Europe and Asia, and as of press time, there also was some traction in the United States. For example, the U.S. Department of Defense (DoD) announced that from October 1, 2003, all new developments and procurements needed to be IPv6-capable; the DoD's goal was to complete the transition to IPv6 for all intra- and internetworking across the agency by 2008. In 2005, the U.S. General Accountability Office (GAO) recommended that all agencies become proactive in planning a coherent transition to IPv6. The expectation is that in the next few years a transition to this new protocol will occur worldwide.

IPv6 is considered to be the next-generation Internet Protocol [HUI199701], [HAG200201], [MUR200501], [SOL200401], [ITO200401], [MIL199701], [MIL200001], [GRA200001], [DAV200201], [LOS200301], [LEE200501], [GON199801], [DEM200301], [GOS200301], [MIN200601], and [WEG199901]. The current version of the Internet Protocol, IPv4, has been in use for almost 30 years and exhibits some challenges in supporting emerging demands for address space cardinality, high-density mobility, multimedia, and strong security. This is particularly true in developing domestic and defense department applications utilizing peer-to-peer networking. IPv6 is an improved version of the Internet Protocol that is designed to coexist with IPv4 and eventually provide better internetworking capabilities than IPv4 [IPV200401].

IPv6 offers the potential of achieving the scalability, reachability, end-to-end interworking, quality of service (QoS), and commercial-grade robustness for data as well as for Voice over IP (VoIP), IP-based TV (IPTV*) distribution, and triple-play networks; these capabilities are mandatory mileposts of the technology if it is to replace the time division multiplexing (TDM) infrastructure around the world.

IPv6 was initially developed in the early 1990s because of the anticipated need for more end-system addresses based on anticipated Internet growth, encompassing mobile phone deployment, smart home appliances, and billions of new users in developing countries (e.g., China and India). New technologies and applications such as VoIP, "always-on access" (e.g., Digital Subscriber Line and cable), Ethernet to the home, converged networks, and evolving ubiquitous computing applications will be driving this need even more in the next few years [IPV200501]. Figure 1.1 depicts one example of a converged network utilizing IPv6, with both local and wide area components as well as private and carrier-provided communications domains; the IPv6/IPv4 network shown in this figure supports video delivery, VoIP, Internet, intranet, and wireless services.

Basic network address translation (NAT) is a method by which IP addresses (specifically IPv4 addresses) are transparently mapped from one group to another. Specifically, private "nonregistered" addresses are mapped to a small set (as small as one) of public registered addresses; this impacts the general addressability, accessibility, and "individuality" of the device. Network address port translation (NAPT) is a method by which many network addresses and their TCP/UDP (Transmission Control Protocol/User Datagram Protocol) ports are translated into a single network address and its TCP/UDP ports. Together, these two methods, referred to as traditional NAT, provide a mechanism to connect a realm with private addresses to an external realm with globally unique registered addresses [RFC3022].

NAT is a short-term solution for the anticipated Internet growth phenomenon, and a better solution is needed for address exhaustion. There is a recognition that NAT techniques make the Internet, the applications, and even the devices more complex; this means a cost overhead [IPV200501]. The expectation is that IPv6 can make IP devices less expensive, more powerful, and even consume less power. The power issue is important not only for environmental reasons, but also for improved operability (e.g., longer battery life in portable devices, such as mobile phones).

Corporations and government agencies will be able to achieve a number of improvements with IPv6. IPv6 can improve a firm's intranet, with benefits such as the following:

- Expanded addressing capabilities
- Server-less autoconfiguration ("plug-and-play") and reconfiguration

---

* IPTV is the delivery of (entertainment-quality) video programming over an IP-based network. Traditional telecom carriers are looking to compete with cable TV companies by deploying IP video services over their networks.

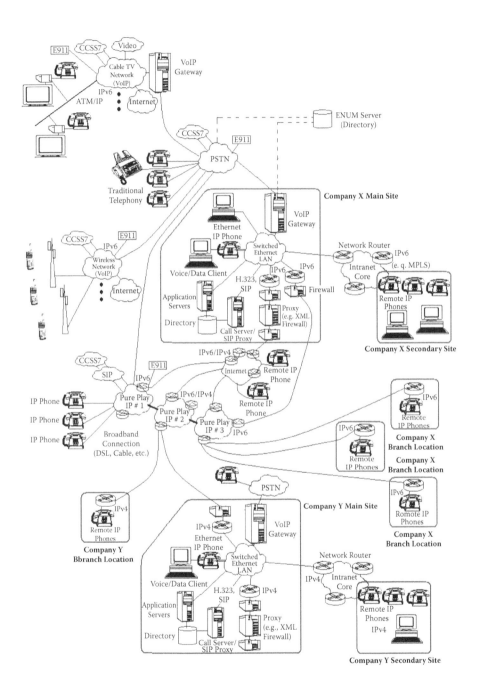

Figure 1.1   Example of the use of IPv6, with emphasis on converged networks.

- More efficient and robust mobility mechanisms
- End-to-end security, with built-in, strong IP-layer encryption and authentication
- Streamlined header format and flow identification
- Enhanced support for multicast and QoS
- Extensibility, with improved support for feature options or extensions

Although the basic function of the Internet Protocol is to move information across networks, IPv6 has more capabilities built into its foundation than IPv4. A key capability is the significant increase in address space. For example, all devices could have a public IP address so that they can be uniquely tracked. Today, inventory management of dispersed IT assets cannot be achieved with IP mechanisms; during the inventory cycle, someone has to manually verify the location of each desktop computer. With IPv6, one can use the network to verify where such equipment is; even non-IT equipment in the field can also be tracked by having an IP address permanently assigned to it. IPv6 also has extensive automatic configuration (autoconfiguration) mechanisms and reduces the IT burden, making configuration essentially plug-and-play.

Corporations and institutions need to start planning the migration process and determining best how coexistence can be maintained during the three- to six-year window that will likely be required to achieve the global worldwide transition. This book addresses the IPv6 migration and macro-level scalability requirements.

1.2 Introductory Overview of IPv6

The Internet Protocol was designed in the 1970s to connect computers that were in separate geographic locations. Computers in a campus were connected by local networks, but these local networks were separated into essentially stand-alone islands. Internet, as a name to designate the protocol and more recently the worldwide information network, simply means *internetwork*, that is, a connection between networks. In the beginning, the protocol had only military use, but computers from universities and enterprises were quickly added. The Internet as a worldwide information network is the result of the practical application of the IP, that is, the result of the interconnection of a large set of information networks [IPV200501]. Starting in the early 1990s, developers realized that the communication needs of the 21st century included a protocol with some new features and capabilities while retaining the useful features of the existing protocol.

Although link-level communication does not generally require a node identifier (address) because the device is intrinsically identified with the link-level address, communication over a group of links (a network) does require unique node identifiers (addresses). The IP address is an identifier that is applied to each device connected to an IP network. In this setup, different elements taking part in the network (servers, routers, user computers, etc.) communicate among each other using their IP address as an entity identifier. In version 4 of the IP, addresses consist of four octets. For ease of human conversation, IP addresses are represented as separated by periods, for example, 166.74.110.83, with the decimal numbers shorthand for (and corresponding to) the binary code described by the byte in question (an eight-bit number takes a value in the 0–255 range). Because the IPv4 address has 32 bits, there are nominally 2^{32} different IP addresses (approximately four billion nodes, if all combinations are used).

IPv6 is the Internet's next-generation protocol, which was at first called IPng (Internet Next Generation). The Internet Engineering Task Force (IETF) developed the basic specifications during the 1990s to support a migration to a new environment. IPv6 is defined in RFC (Request for Comment) 2460, Internet Protocol, Version 6 (IPv6) Specification, by S. Deering and R. Hinden (December 1998) [RFC2460], which makes obsolete RFC 1883. (IP version 5 was employed for another use, an experimental real-time streaming protocol, and to avoid any confusion, it was decided not to use this nomenclature.)

1.2.1 IPv6 Benefits

IPv4 has proven, by means of its long life, to be a flexible and powerful networking mechanism. However, IPv4 is starting to exhibit limitations, not only with respect to the need for an increase of the IP address space, driven, for example, by new populations of users in countries like China and India and by new technologies with always-connected devices (DSL, cable, networked personal digital assistants [PDAs], 2.5G/3G mobile telephones, etc.), but also in reference to a potential global rollout of VoIP. IPv6 creates a new IP address format so that the number of IP addresses will not be exhausted for several decades or longer even though an entire new crop of devices is expected to connect to Internet.

IPv6 also adds improvements in areas such as routing and network autoconfiguration. Specifically, new devices that connect to Internet will be plug-and-play devices. With IPv6, one is not required to configure dynamic nonpublished local IP addresses, the gateway address, the subnetwork mask, or any other parameters. The equipment, when plugged into the network, automatically obtains all requisite configuration data [IPV200501].

The advantages of IPv6 can be summarized as follows:

■ Scalability: IPv6 has 128-bit addresses versus 32-bit IPv4 addresses. With IPv4, the theoretical number of available IP addresses is $2^{32} \sim 10^{10}$. IPv6 offers a 2^{128} space. Hence, the number of available unique node addressees is $2^{128} \sim 10^{39}$.
■ Security: IPv6 includes security features, such as payload encryption and authentication of the source of the communication, in its specifications.

- Real-time applications: To provide better support for real-time traffic (e.g., VoIP), IPv6 includes "labeled flows" in its specifications. By means of this mechanism, routers can recognize the end-to-end flow to which transmitted packets belong. This is similar to the service offered by Multi-Protocol Label Switching (MPLS), but it is intrinsic with the IP mechanism rather than an add-on. Also, it preceded this MPLS feature by a number of years.
- Plug-and-play: IPv6 includes a plug-and-play mechanism that facilitates the connection of equipment to the network. The requisite configuration is automatic.
- Mobility: IPv6 includes more efficient and enhanced mobility mechanisms, particularly important for mobile networks.
- Optimized protocol: IPv6 embodies IPv4 best practices but removes unused or obsolete IPv4 characteristics. This results in a better-optimized Internet Protocol.
- Addressing and routing: IPv6 improves the addressing and routing hierarchy.
- Extensibility: IPv6 has been designed to be extensible and offers support for new options and extensions.

1.2.2 Traditional Addressing Classes for IPv4

With IPv4, the 32-bit address can be represented as AdrClass|netID|hostID. The network portion can contain either a network ID or a network ID and a subnet. Every network and every host or device has a unique address by definition. Figure 1.2 depicts the traditional address classes.

- Traditional Class A Address. Class A uses the first bit of the 32-bit space (bit 0) to identify it as a Class A address; this bit is set to 0. Bits 1 to 7 represent the network ID, and bits 8 through 31 identify the personal computer (PC), terminal device, or host/server on the network. This address space supports $2^7 - 2 = 126$ networks and approximately 16 million devices (2^{24}) on each network. By convention, the use of an "all 1s" or "all 0s" address for both the network ID and the host ID is prohibited (which is the reason for subtracting the 2 in the equation).
- Traditional Class B Address. Class B uses the first 2 bits (bit 0 and bit 1) of the 32-bit space to identify it as a Class B address; these bits are set to 10. Bits 2 to 15 represent the network ID, and bits 16 through 31 identify the PC, terminal device, or host/server on the network. This address space supports $2^{14} - 2 = 16,382$ networks and $2^{16} - 2 = 65,134$ devices on each network.
- Traditional Class C Address. Class C uses the first 3 bits (bit 0, bit 1, and bit 2) of the 32-bit space to identify it as a Class C address; these bits are set

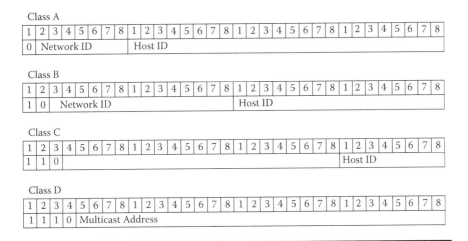

Figure 1.2 Traditional address classes for IP address.

to 110. Bits 3 to 23 represent the network ID, and bits 24 through 31 identify the PC, terminal device, or host/server on the network. This address space supports about 2 million networks ($2^{21} - 2$) and $2^8 - 2 = 254$ devices on each network.

■ Traditional Class D Address. This class is used for broadcasting in which all devices on the network receive the same packet. Class D uses the first 4 bits (bit 0, bit 1, bit 2, and bit 3) of the 32-bit space to identify it as a Class D address; these bits are set to 1110. This is used in IP Multicast applications (for example for IPTV).

Classless interdomain routing (CIDR), described in RFC 1518, RFC 1519, and RFC 2050, is yet another mechanism that was developed to help alleviate the problem of exhaustion of IP addresses and growth of routing tables. The concept behind CIDR is that blocks of multiple addresses (e.g., blocks of Class C addresses) can be combined, or aggregated, to create a larger classless set of IP addresses with more hosts allowed. Blocks of Class C network numbers are allocated to each network service provider; organizations using the network service provider for Internet connectivity are allocated subsets of the service provider's address space as required. These multiple Class C addresses can then be summarized in routing tables, resulting in fewer route advertisements. The CIDR mechanism can also be applied to blocks of Class A and B addresses [TEA200401]. All of this assumes, however, that the institution in question already has an assigned set of public, registered IP addresses; it does not address the issue of how to get additional public, registered globally unique IP addresses.

1.2.3 Network Address Translation Issues in IPv4

IPv4 addresses can be from an officially assigned public range or from an internal intranet private (but not globally unique) block. Internal intranet addresses may be in the ranges 10.0.0.0/8, 172.16.0.0/12, and 192.168.0.0/16. In the internal intranet private address case, an NAT function is employed to map the internal addresses to an external public address when the private-to-public network boundary is crossed. This imposes a number of limitations, particularly because the number of registered public addresses available to a company is almost invariably much smaller (as small as one) than the number of internal devices requiring an address.

As noted, IPv4 theoretically allows up to 2^{32} addresses, based on a four-octet address space. Public, globally unique addresses are assigned by the Internet Assigned Numbers Authority (IANA). IP addresses are addresses of network nodes at Layer 3; each device on a network (whether the Internet or an intranet) must have a unique address. In IPv4, it is a 32-bit (four-byte) binary address used to identify the device. It is represented by the nomenclature a.b.c.d (with each of a, b, c, and d from 1 to 255; 0 has a special meaning). Examples are 167.168.169.170, 232.233.229.209, and 200.100.200.100.

The problem is that during the 1980s many public, registered addresses were allocated to firms and organizations without any consistent control. As a result, some organizations have more addresses than they actually need, giving rise to the present dearth of available "registerable" Layer 3 addresses. Furthermore, not all IP addresses can be used due to the fragmentation described.

One approach to the issue would be renumbering and reallocation of the IPv4 addressing space. However, this is not as simple as it appears because it requires worldwide coordination efforts. Moreover, it would still be limited for the human population and the quantity of devices that will be connected to the Internet in the medium-term future [IPV200501]. At this juncture, and as a temporary and pragmatic approach to alleviate the dearth of addresses, NAT mechanisms are employed by organizations and even home users. This mechanism consists of using only a small set of public IPv4 addresses for an entire network to access the Internet. The myriad internal devices are assigned IP addresses from a specifically designated range of Class A or Class C addresses that are locally unique but are duplicatively used and reused within various organizations. In some cases (e.g., residential Internet access use via DSL or cable), the legal IP address is only provided to a user on a time-lease basis rather than permanently.

A number of protocols cannot travel through an NAT device, and hence the use of NAT implies that many applications (e.g., VoIP) cannot be used effectively in all instances. As a consequence, these applications can only be used in intranets. Examples include [IPV200501] the following:

■ Multimedia applications such as videoconferencing, VoIP, or video-on-demand/IPTV do not work smoothly through NAT devices. Multimedia applications make use of Real-Time Transport Protocol (RTP) and Real-Time Control Protocol (RTCP). These in turn use UDP with dynamic allocation of ports, and NAT does not directly support this environment.

- Kerberos authentication needs the source address, and the source address in the IP header is often modified by NAT devices.
- IPSec is used extensively for data authentication, integrity, and confidentiality. However, when NAT is used, there is an impact on IPSec because NAT changes the address in the IP header.
- Multicast, although possible in theory, requires complex configuration in a NAT environment and hence in practice is not utilized as often as could be the case.

The need for obligatory use of NAT disappears with IPv6.

1.2.4 IPv6 Address Space

The format of IPv6 addressing is described in RFC 2373. As noted, an IPv6 address consists of 128 bits rather than 32 bits as with IPv4 addresses. The number of bits correlates to the address space as follows:

IP Version	Size of Address Space
IPv6	128 bits, which allows for 2^{128} or 340,282,366,920,938,463,463,374,607,431,768,211,456 (3.4 (10^{38}) possible addresses
IPv4	32 bits, which allows for 2^{32} or 4,294,967,296 possible addresses

The relatively large size of the IPv6 address is designed to be subdivided into hierarchical routing domains that reflect the topology of the modern-day Internet. The use of 128 bits provides multiple levels of hierarchy and flexibility in designing hierarchical addressing and routing. The IPv4-based Internet currently lacks this flexibility [MSD200401].

The IPv6 address is represented as eight groups of 16 bits each, separated by the ":" character. Each 16-bit group is represented by four hexadecimal digits; that is, each digit has a value between 0 and F (0, 1, 2, ... A, B, C, D, E, F with A = 10, B = 11, and so on to F = 15). What follows is an IPv6 address example:

3223:0BA0:01E0:D001:0000:0000:D0F0:0010

If one or more four-digit groups is 0000, the zeros may be omitted and replaced with colons (::). For example,

3223:0BA0::

is the abbreviated form of the following address:

3223:0BA0:0000:0000:0000:0000:0000:0000

Similarly, the address

3223:0BA0::1234

is the abbreviated form of the following address:

3223:0BA0:0000:0000:0000:0000:0000:1234

There is also a method to designate groups of IP addresses or subnetworks that is based on specifying the number of bits that designate the subnetwork, beginning from left to right, using remaining bits to designate single devices inside the network. For example, the notation

3223:0BA0:01A0::/48

indicates that the part of the IP address used to represent the subnetwork has 48 bits. Because each hexadecimal digit has four bits, this points out that the part used to represent the subnetwork is formed by 12 digits, that is: 3223:0ba0:01a0. The remaining digits of the IP address would be used to represent nodes inside the network.

There are a number of special IPv6 addresses:

■ Autoreturn or loopback virtual address. This address is specified in IPv4 as the 127.0.0.1 address. In IPv6, this address is represented as ::1.
■ Unspecified address (::). This address is not allocated to any node because it is used to indicate the absence of an address.
■ IPv6 over IPv4 dynamic/automatic tunnel addresses. These addresses are designated as IPv4-compatible IPv6 addresses and allow the sending of IPv6 traffic over IPv4 networks in a transparent manner. They are represented as, for example, ::156.55.23.5.
■ IPv4 over IPv6 addresses automatic representation. These addresses allow for IPv4-only nodes to still work in IPv6 networks. They are designated as IPv4-mapped IPv6 addresses and are represented as ::FFFF:, for example, ::FFFF.156.55.43.3.

1.2.5 Basic Protocol Constructs

Table 1.1 lists basic IPv6 terminology (see Appendix A for a more inclusive glossary). Table 1.2 shows the core protocols that comprise IPv6 (see Appendix B for a more inclusive listing).

Like IPv4, IPv6 is a connectionless, unreliable datagram protocol used primarily for addressing and routing packets between hosts. *Connectionless* means that a session is not established before exchanging data. *Unreliable* means that delivery is not guaranteed. IPv6 always makes a best-effort attempt to deliver a packet. An IPv6 packet might be lost, delivered out of sequence, duplicated, or delayed. IPv6

Table 1.1   Basic IPv6 Terminology

Address	An IP layer identifier for an interface or a set of interfaces.
Host	Any node that is not a router.
Interface	A node's attachment to a link.
Link	A communication facility or medium over which nodes can communicate at the link layer, that is, the layer immediately below IP. Examples are Ethernet (simple or bridged); PPP links, X.25, Frame Relay, ATM networks and Internet (or higher) layer tunnels, such as tunnels over IPv4 or IPv6 itself.
Link-layer identifier	A link-layer identifier for an interface. Examples include IEEE 802 addresses for Ethernet network interfaces and E.164 addresses for Integrated Services Digital Network (ISDN) links.
Link-local address	An IPv6 address having a link-only scope, indicated by having the prefix (FE80::/10), that can be used to reach neighboring nodes attached to the same link. Every interface has a link-local address.
Multicast address	An identifier for a set of interfaces typically belonging to different nodes. A packet sent to a multicast address is delivered to all interfaces identified by that address.
Neighbor	A node attached to the same link.
Node	A device that implements IP.
Packet	An IP header plus payload.
Prefix	The initial bits of an address or a set of IP addresses that share the same initial bits.
Prefix length	The number of bits in a prefix.
Router	A node that forwards IP packets not explicitly addressed to itself.
Unicast address	An identifier for a single interface. A packet sent to a unicast address is delivered to the interface identified by that address.

per se does not attempt to recover from these types of errors. The acknowledgment of packets delivered and the recovery of lost packets is done by a higher-layer protocol, such as TCP [MSD200401]. From a packet-forwarding perspective, IPv6 operates just like IPv4. An IPv6 packet, also known as an IPv6 datagram, consists of an IPv6 header and an IPv6 payload, as shown Figure 1.3.

Table 1.2   Key IPv6 Protocols

Protocol	Description
Internet Protocol version 6 (IPv6): RFC 2460	IPv6 is a connectionless datagram protocol used for routing packets between hosts.
Internet Control Message Protocol for IPv6 (ICMPv6): RFC 2463	ICMPv6 is a mechanism that enables hosts and routers that use IPv6 communication to report errors and send simple status messages.
Multicast Listener Discovery (MLD): RFC 2710, RFC 3590, RFC 3810	MLD is a mechanism that enables one to manage subnet multicast membership for IPv6. MLD uses a series of three ICMPv6 messages. MLD replaces the Internet Group Management Protocol (IGMP) version 3 that is employed for IPv4.
Neighbor Discovery (ND): RFC 2461	ND is a mechanism that is used to manage node-to-node communication on a link. ND uses a series of five ICMPv6 messages. ND replaces Address Resolution Protocol (ARP), ICMPv4 Router Discovery, and the ICMPv4 Redirect message. It also provides additional functions.

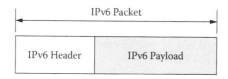

Figure 1.3   IPv6 packet.

The IPv6 header consists of two parts: the IPv6 base header and optional extension headers. Functionally, the optional extension headers and upper-layer protocols (e.g., TCP) are considered part of the IPv6 payload. Table 1.3 shows the fields in the IPv6 base header. IPv4 and IPv6 headers are not directly interoperable: Hosts or routers must use an implementation of both IPv4 and IPv6 to recognize and process both header formats. This gives rise to a number of complexities in the migration process between the IPv4 and the IPv6 environments. However, techniques have been developed to handle these migrations, as we discuss later in the book.

Table 1.3 IPv6 Base Header

IPv6 Header Field	Length (bits)	Function
Version	4	Identifies the version of the protocol. For IPv6, the version is 6.
Traffic Class	8	Intended for originating nodes and forwarding routers to identify and distinguish between different classes or priorities of IPv6 packets.
Flow Label	20	(Sometimes referred to as flow ID.) Defines how traffic is handled and identified. A flow is a sequence of packets sent either to a unicast or a multicast destination. This field identifies packets that require special handling by the IPv6 node. The following list shows the ways the field is handled if a host or router does not support flow label field functions: • If the packet is being sent, the field is set to zero. • If the packet is being received, the field is ignored.
Payload Length	16	Identifies the length, in octets, of the payload. This field is a 16-bit unsigned integer. The payload includes the optional extension headers as well as the upper-layer protocols, for example, TCP.
Next Header	8	Identifies the header immediately following the IPv6 header. The following are examples of the next header: • 00 = Hop-by-Hop options • 01 = ICMPv4 • 04 = IP in IP (encapsulation) • 06 = TCP • 17 = UDP • 43 = Routing • 44 = Fragment • 50 = Encapsulating security payload • 51 = Authentication • 58 = ICMPv6

Table 1.3 IPv6 Base Header (continued)

IPv6 Header Field	Length (bits)	Function
Hop Limit	8	Identifies the number of network segments, also known as links or subnets, on which the packet is allowed to travel before being discarded by a router. The hop limit is set by the sending host and is used to prevent packets from endlessly circulating on an IPv6 internetwork.
		When forwarding an IPv6 packet, IPv6 routers must decrease the hop limit by 1 and must discard the IPv6 packet when the hop limit is 0.
Source Address	128	Identifies the IPv6 address of the original source of the IPv6 packet.
Destination Address	128	Identifies the IPv6 address of the intermediate or final destination of the IPv6 packet.

1.2.6 IPv6 Autoconfiguration

Autoconfiguration is a new characteristic of IPv6 that facilitates network management and system setup tasks by users. This characteristic is often called "plug-and-play" or "connect-and-work." Autoconfiguration facilitates initialization of user devices: After connecting a device to an IPv6 network, one or several IPv6 globally unique addresses are automatically allocated.

The autoconfiguration process is flexible, but it is also somewhat complex. The complexity arises from the fact that various policies are defined and implemented by the network administrator. Specifically, the administrator determines the parameters that will be assigned automatically. At a minimum (or when there is no network administrator), the allocation of a "link-local" address is often included. The link-local address allows communication with other nodes placed in the same physical network. Note that the word *link* has somewhat of a special meaning in IPv6; it indicates a communication facility or medium over which nodes can communicate at the link layer, that is, the layer immediately below IPv6. Examples are Ethernets (simple or bridged); PPP (Point-to-Point Protocol) links; an X.25 packet-switched network; a Frame Relay network; Cell Relay/Asynchronous Transfer Mode (ATM) network; and Internet(working) layer (or higher layer) tunnels, such as tunnels over IPv4 or IPv6 itself [RFC2460].

Two autoconfiguration basic mechanisms exist: stateful and stateless. Both mechanisms can be used in a complementary manner or simultaneously to define parameter configurations [IPV200501].

Stateless autoconfiguration is also described as "serverless." Here, the presence of configuration servers to supply profile information is not required. The host generates its own address using a combination of the information that it possesses (in its interface or network card) and the information that is periodically supplied by the routers. Routers determine the prefix that identifies networks associated to the link under discussion. The *interface identifier* identifies an interface within a subnetwork and is often, and by default, generated from the Media Access Control (MAC) address of the network card. The IPv6 address is built combining the 64 bits of the interface identifier with the prefixes that routers determine as belonging to the subnetwork. If there is no router, then the interface identifier is self-sufficient to allow the PC to generate a link-local address. The link-local address is sufficient to allow the communication between several nodes connected to the same link (the same local network).

Stateful configuration requires a server to send the information and parameters of network connectivity to nodes and hosts. Servers maintain a database with all addresses allocated and a mapping of the hosts to which these addresses have been allocated, along with any information related to all requisite parameters. In general, this mechanism is based on the use of Dynamic Host Control Protocol version 6 (DHCPv6).

Stateful autoconfiguration is often employed when there is a need for rigorous control in reference to the address allocated to hosts; in stateless autoconfiguration, the only concern is that the address be unique. Depending on the network administrator policies, it may be required that some addresses be allocated to specific hosts and devices in a permanent manner; here, the stateful mechanism is employed on this subset of hosts, but the control of the remaining parameters or nodes could be less rigorous. In some environments, there are no policy requirements on the importance of the allocated addresses, but there may be rules on the parameters, such as that they be allocated in a certain "static" manner, with information stored in a server.

IPv6 addresses are "leased" to an interface for a fixed established time (including an infinite time). When this "lifetime" expires, the link between the interface and the address is invalidated, and the address can be reallocated to other interfaces. For the suitable management of address expiration time, an address goes through two states (stages) when it is affiliated to an interface [IPV200501]:

1. At first, an address is in a "preferred" state, so its use in any communication is not restricted.
2. After that, an address becomes "deprecated," indicating that its affiliation with the current interface will (soon) be invalidated.

When in a deprecated state, the use of the address is discouraged, although it is not forbidden. However, when possible, any new communication (e.g., the opening of a new TCP connection) must use a preferred address. A deprecated address should only be used by applications that already used it before and if it is difficult to change this address to another address without causing a service interruption.

To ensure that allocated addresses (granted either by manual mechanisms or by autoconfiguration) are unique in a specific link, the link duplicated addresses detection algorithm is used. The address to which the duplicated address detection algorithm is applied is designated (until the end of this algorithmic session) as an "attempt address." In this case, it does not matter that such address has been allocated to an interface and that received packets are discarded.

Next, we describe how an IPv6 address is formed. The lowest 64 bits of the address identify a specific interface, and these bits are designated as "interface identifier." The highest 64 bits of the address identify the "path" or the "prefix" of the network or router in one of the links to which such interface is connected. The IPv6 address is formed by combining the prefix with the interface identifier.

It is possible for a host or device to have IPv6 and IPv4 addresses simultaneously. Most of the systems that currently support IPv6 allow the simultaneous use of both protocols. In this way, it is possible to support communication with IPv4-only networks as well as IPv6-only networks and the use of the applications developed for both protocols [IPV200501].

It is possible to transmit IPv6 traffic over IPv4 networks via tunneling methods. This approach consists of "wrapping" the IPv6 traffic as IPv4 payload data: IPv6 traffic is sent "encapsulated" into IPv4 traffic, and at the receiving end this traffic is parsed as IPv6 traffic. Transition mechanisms are methods used for the coexistence of IPv4 or IPv6 devices and networks. For example, an IPv6-in-IPv4 tunnel is a transition mechanism that allows IPv6 devices to communicate through an IPv4 network. The mechanism consists of creating the IPv6 packets in a normal way and encapsulating them in an IPv4 packet. The reverse process is undertaken in the destination machine, which deencapsulates the IPv6 packet.

There is a significant difference between the procedures to allocate IPv4 addresses, which focus on the parsimonious use of addresses (because addresses are a scarce resource and should be managed with caution), and the procedures to allocate IPv6 addresses, which focus on flexibility. Internet Service Providers (ISPs) deploying IPv6 systems follow the Regional Internet Registries (RIRs) policies relating to how to assign IPv6 addressing space among their clients. RIRs are recommending ISPs and operators allocate to each IPv6 client a /48 subnetwork; this allows clients to manage their own subnetworks without using NAT. (The implication is that the need for NAT disappears in IPv6.)

To allow its maximum scalability, IPv6 uses an approach based on a basic header, with minimum information. This differentiates it from IPv4, in which different options are included in addition to the basic header. IPv6 uses a header "concatenation" mechanism to support supplementary capabilities. The advantages of this approach include the following:

■ The size of the basic header is always the same and is well known. The basic header has been simplified compared with IPv4 because only eight fields are used instead of twelve. The basic IPv6 header has a fixed size; hence, its processing by nodes and routers is more straightforward. Also, the header's structure aligns to 64 bits, so that new and future processors (64 bits minimum) can process it in a more efficient way.

■ Routers placed between a source point and a destination point (that is, the route that a specific packet has to pass through) do not need to process or understand any "following headers." In other words, in general, interior (core) points of the network (routers) only have to process the basic header; in IPv4, all headers must be processed. This flow mechanism is similar to the operation in MPLS yet precedes it by several years.

■ There is no limit to the number of options that the headers can support (the IPv6 basic header is 40 octets in length; the IPv4 one varies from 20 to 60 octets, depending on the options used).

In IPv6, interior/core routers do not perform packet fragmentation, but the fragmentation is performed end to end. That is, source and destination nodes perform, by means of the IPv6 stack, the fragmentation of a packet and the reassembly, respectively. The fragmentation process consists of dividing the source packet into smaller packets or fragments [IPV200501].

A *jumbogram* is an option that allows an IPv6 packet to have a payload greater than 65,535 bytes. Jumbograms are identified with a 0 value in the payload length in the IPv6 header field and include a Jumbo Payload Option in the Hop-by-Hop Option header. It is anticipated that such packets will be used in particular for multimedia traffic.

This preliminary overview of IPv6 highlights the advantages of the new protocol and its applicability to a whole range of applications, including VoIP.

1.3 Migration and Coexistence

Migration is expected to be fairly complex. Initially, internetworking between the two environments will be critical. Existing IPv4 endpoints or nodes will need to run dual-stack nodes or convert to IPv6 systems. Fortunately the new protocol supports IPv4-compatible IPv6 addresses, which is an IPv6 address format that employs embedded IPv4 addresses. Tunneling, which we already described in passing, will play a major role in the beginning.

There are a number of requirements that are typically applicable to an organization wishing to introduce an IPv6 service [6NE200501]:

- The existing IPv4 service should not be adversely disrupted (e.g., as it might be by router loading of encapsulating IPv6 in IPv4 for tunnels).
- The IPv6 service should perform as well as the IPv4 service (e.g., at the IPv4 line rate and with similar network characteristics).
- The service must be manageable and be able to be monitored (thus tools should be available for IPv6 as they are for IPv4).
- The security of the network should not be compromised due to the additional protocol itself or a weakness of any transition mechanism used.
- An IPv6 address allocation plan must be drawn up.

Well-known interworking mechanisms include the following [GIL200001]:

- Dual IP layer (also known as dual stack): A technique for providing complete support for both Internet Protocols — IPv4 and IPv6 — in hosts and routers
- Configured tunneling of IPv6 over IPv4: Point-to-point tunnels made by encapsulating IPv6 packets within IPv4 headers to carry them over IPv4 routing infrastructures
- Automatic tunneling of IPv6 over IPv4: A mechanism for using IPv4-compatible addresses to automatically tunnel IPv6 packets over IPv4 networks

Tunneling techniques include the following [RFC2893]:

- IPv6 over IPv4 tunneling: The technique of encapsulating IPv6 packets within IPv4 so that they can be carried across IPv4 routing infrastructures.
- Configured tunneling: IPv6 over IPv4 tunneling in which the IPv4 tunnel endpoint address is determined by configuration information on the encapsulating node. The tunnels can be either unidirectional or bidirectional. Bidirectional configured tunnels behave as virtual point-to-point links.
- Automatic tunneling: IPv6 over IPv4 tunneling in which the IPv4 tunnel endpoint address is determined from the IPv4 address embedded in the IPv4-compatible destination address of the IPv6 packet being tunneled.
- IPv4 multicast tunneling: IPv6 over IPv4 tunneling in which the IPv4 tunnel endpoint address is determined using Neighbor Discovery. Unlike configured tunneling, this does not require any address configuration, and unlike automatic tunneling, it does not require the use of IPv4-compatible addresses. However, the mechanism assumes that the IPv4 infrastructure supports IPv4 multicast.

Applications (and the lower-layer protocol stack) need to be properly equipped. There are four cases [SHI200501]:

Case 1: IPv4-only applications in a dual-stack node. IPv6 is introduced in a node, but applications are not yet ported to support IPv6. The protocol stack is as follows:

```
+------------------------+
|          appv4         |    (appv4 - IPv4-only applications)
+------------------------+
|   TCP / UDP / others   |    (transport protocols - TCP, UDP,
+------------------------+     SCTP, DCCP, etc.)
|      IPv4 | IPv6       |    (IP protocols supported/enabled in the OS)
+------------------------+
```

Case 2: IPv4-only applications and IPv6-only applications in a dual-stack node. Applications are ported for IPv6 only. Therefore, there are two similar applications, one for each protocol version (e.g., ping and ping6). The protocol stack is as follows:

```
+------------------------+    (appv4 - IPv4-only applications)
|    appv4 | appv6      |    (appv6 - IPv6-only applications)
+------------------------+
|   TCP / UDP / others   |    (transport protocols - TCP, UDP, SCTP,
+------------------------+     DCCP, etc.)
|      IPv4 | IPv6       |    (IP protocols supported/enabled in the OS)
+------------------------+
```

Case 3: Applications supporting both IPv4 and IPv6 in a dual-stack node. Applications are ported for both IPv4 and IPv6 support. Therefore, the existing IPv4 applications can be removed. The protocol stack is as follows:

```
+-------------------------+
|         appv4/v6        |    (appv4/v6 - applications supporting both
+-------------------------+     IPv4 and IPv6)
|   TCP / UDP / others    |    (transport protocols - TCP, UDP, SCTP,
+-------------------------+     DCCP, etc.)
|      IPv4 | IPv6        |    (IP protocols supported/enabled in the OS)
+-------------------------+
```

Case 4: Applications supporting both IPv4 and IPv6 in an IPv4-only node. Applications are ported for both IPv4 and IPv6 support, but the same applications may also have to work when IPv6 is not in use (e.g., disabled from the OS). The protocol stack is as follows:

```
+---------------------------+
|          appv4/v6         |    (appv4/v6 - applications supporting both
+---------------------------+     IPv4 and IPv6)
|     TCP / UDP / others    |    (transport protocols - TCP, UDP, SCTP,
+---------------------------+     DCCP, etc.)
|           IPv4            |    (IP protocols supported/enabled in the OS)
+---------------------------+
```

The first two cases are not interesting in the longer term; only a few applications are inherently IPv4 or IPv6 specific and should work with both protocols without having to care about which one is used.

With transition, migration, and coexistence the major topics of this text, we defer detailed discussion to the chapters that follow.

1.4 Course of Investigation

After this introduction, we provide a quick tutorial of the addressing capabilities in Chapter 2; Chapter 3 looks at IPv6 network constructs, specifically key routing processes; Chapter 4 examines the IPv6 autoconfiguration techniques. To wrap up this portion of the text, Chapter 5 looks more formally at the suite of IPv6 and related protocols.

Chapter 6 starts the discussion of the major theme of this text: IPv6 enterprise/institutional network migration scenarios (tunneling and encapsulation) and coexistence issues. Chapter 7 looks at the various elements in the network and which migration role they need to play to support the transition. Chapter 8 presents an overview of example transition strategies. Chapter 9 concerns application aspects of IPv6 transition. Chapter 10 concludes the discussion by looking at security in IPv6 networks.

References

[6NE200501] 6NET, D2.2.4: Final IPv4 to IPv6 Transition Cookbook for Organisational/ISP (NREN) and Backbone Networks, version 1.0 (February 4, 2005), Project Number IST-2001-32603, CEC Deliverable Number 32603/UOS/DS/2.2.4/A1.

[DAV200201] J. Davies, *Understanding IPv6*, Microsoft Press, 2002.

[DEM200301] R. Desmeules, *Cisco Self-Study: Implementing IPv6 Networks (IPV6)*, Pearson Education, May 2003.

[GON199801] M. Goncalves and K. Niles, *IPv6 Networks*, McGraw-Hill Osborne, 1998.

[GOS200301] S. Goswami, *Internet Protocols: Advances, Technologies, and Applications*, Kluwer Academic Publishers, May 2003.

[GRA200001] B. Graham, *TCP/IP Addressing: Designing and Optimizing Your IP Addressing Scheme* (2nd edition), Morgan Kaufmann, 2000.

[HAG200201] S. Hagen, *IPv6 Essentials*, O'Reilly, 2002.

[HUI199701] C. Huitema, *IPv6 the New Internet Protocol* (2nd edition), Prentice Hall, 1997.

[IPV200401] IPv6Forum, IPv6 Vendors Test Voice, Wireless and Firewalls on Moonv6, November 15, 2004, http://www.ipv6forum.com/modules.php?op=modload&name =News&file=article&sid=15&mode=thread&order=0&thold=0.

[IPV200501] IPv6 Portal, http://www.ipv6tf.org/meet/faqs.php.

[ITO200401] J. Itojun Hagino, *IPv6 Network Programming*, Butterworth-Heinemann, 2004.

[LEE200501] H. K. Lee, *Understanding IPv6*, Springer-Verlag, New York, 2005.

[LOS200301] P. Loshin, *IPv6: Theory, Protocol, and Practice* (2nd edition), Elsevier Science and Technology Books, 2003.

[MIL199701] M. A. Miller, *Implementing IPv6: Migrating to the Next Generation Internet Protocol*, John Wiley and Sons, 1997.

[MIL200001] M. Miller and P. E. Miller, *Implementing IPV6: Supporting the Next Generation Internet Protocols* (2nd edition), Hungry Minds, 2000.

[MIN200601] D. Minoli, *VoIP over IPv6*, Elsevier, 2006.

[MSD200401] Microsoft Corporation, MSDN Library, Internet Protocol, 2004, http://msdn.microsoft.com.

[MUR200501] N. R. Murphy and D. Malone, *IPv6 Network Administration*, O'Reilly and Associates, 2005.

[RFC2460] S. Deering and R. Hinden, Internet Protocol, Version 6 (IPv6) Specification, RFC 2460, December 1998.

[RFC2893] R. Gilligan and E. Nordmark, Transition Mechanisms for IPv6 Hosts and Routers, RFC 2893, August 2000.

[RFC3022] P. Srisuresh and K. Egevang Traditional IP Network Address Translator (Traditional NAT), RFC 3022, January 2001.

[SHI200501] M.-K. Shin, Ed., Y.-G. Hong, J. Hagino, P. Savola, and E. M. Castro, Application Aspects of IPv6 Transition, RFC 4038, March 2005.

[SOL200401] H. S. Soliman, *Mobile IPv6*, Pearson Education, 2004.

[TEA200401] D. Teare and C. Paquet, CCNP Self-Study: Advanced IP Addressing, Cisco Press, June 11, 2004.

[WEG199901] J. D. Wegner, *IP Addressing and Subnetting, Including IPv6*, Elsevier Science and Technology Books, 1999.

Chapter 2

IPv6 Addressing

2.1 Introduction

This chapter covers the Internet Protocol version 6 (IPv6) addressing scheme in some detail. Chapter 1 introduced some basic concepts of addressing, and these concepts are expanded in this chapter. A more complete and formal treatment of the topic is provided in Chapter 5.

The IPv6 addressing scheme is defined in the IPv6 Addressing Architecture specification, Internet Engineering Task Force (IETF) Request for Comment (RFC) 4291, April 2003 [RFC4291] (RFC 4291 obsoletes RFC 2373). The IPv6 Addressing Architecture specification defines the address scope that can be used in an IPv6 implementation and the various configuration architecture guidelines for network designers of the IPv6 address space. Two advantages of IPv6 are that support for multicast is intrinsic (it is required by the specification), and nodes can create link-local addresses during initialization [RFC3315]. Some portions of this discussion are based on [MSD200401].

2.2 IPv6 Addressing Mechanisms

2.2.1 Addressing Conventions

As we saw in Chapter 1, the IPv6 128-bit address is divided along 16-bit boundaries; each 16-bit block is then converted to a four-digit hexadecimal number, separated by colons. The resulting representation is called colon-hexadecimal. This is in contrast to the 32-bit IPv4 address represented in dotted-decimal format, divided along 8-bit boundaries, and then converted to its decimal equivalent, separated by periods. The following examples show 128-bit IPv6 addresses in binary form:

Address 1: 0010000111011010000000000110100110000000000000000010
11110011101100000010101010100000000011111111111111110001010
001001110001011010

Address 2: 0010000111011010000000000110100110000000000000000010
11110011101100000010101010100000000011111111100000000000000
001001110001011010

Address 3: 0010000111011010000000000110100110000000000000001001
11000101101000000010101010100000000011111111100000000000000
001001110001011010

Address 4: 0010000111011010000000000110100110000000000000001001
11000101101000000010101010100000000011111111100000000000000
000010111100111011

The following example shows these same addresses divided along 16-bit boundaries:

Address 1: 0010000111011010:0000000011010011:0000000000000000:001
0111100111011:0000001010101010:0000000011111111:1111111100010
1000:1001110001011010:

Address 2: 0010000111011010:0000000011010011:0000000000000000:001
0111100111011:0000001010101010:0000000011111111:000000000000
0000:1001110001011010:

Address 3: 0010000111011010:0000000011010011:0000000000000000:100
1110001011010:0000001010101010:0000000011111111:000000000000
0000:1001110001011010:

Address 4: 0010000111011010:0000000011010011:0000000000000000:100
1110001011010:0000001010101010:0000000011111111:000000000000
0000:0010111100111011:

The following shows each 16-bit block in the address converted to hexadecimal and delimited with colons.

Address 1: 21DA:00D3:0000:2F3B:02AA:00FF:FE28:9C5A
Address 2: 21DA:00D3:0000:2F3B:02AA:00FF:0000:9C5A
Address 3: 21DA:00D3:0000: 9C5A:02AA:00FF:0000:9C5A
Address 4: 21DA:00D3:0000: 9C5A:02AA:00FF:0000:2F3B

IPv6 representations can be further simplified by removing the *leading* zeros (trailing zeros are not removed) within each 16-bit block. However, each block must have, in the abbreviated nomenclature, at least a single digit. The following example shows the addresses without the *leading* zeros:

Address 1: 21DA:D3:0:2F3B:2AA:FF:FE28:9C5A
Address 2: 21DA:D3:0:2F3B:2AA:FF:0:9C5A
Address 3: 21DA:D3:0: 9C5A:2AA:FF:0:9C5A
Address 4: 21DA:D3:0: 9C5A:2AA:FF:0:2F3B

Some types of addresses contain long sequences of zeros. In IPv6 addressing, a contiguous sequence of 16-bit blocks set to 0 in the colon-hexadecimal format can be compressed to :: (known as *double-colon*).

The following list shows examples of compressing zeros:

- The address 21DA:0:0:0:2AA:FF:9C5A:2F3B can be compressed to 21DA:: 2AA:FF:9C5A:2F3B.
- The multicast address of FF02:0:0:0:0:0:0:2 can be compressed to FF02::2.

Note that zero compression can only be used to compress a single contiguous series of 16-bit blocks expressed in colon-hexadecimal notation — one cannot use zero compression to include part of a 16-bit block (e.g., one *cannot* abbreviate FF01:30:0:0:0:0:0:8 as FF01:3::8). Also, zero compression can be used only once in an address, which enables one to determine the number of 0 bits represented by each instance of a double-colon (::). To determine how many 0 bits are represented by the ::, one can count the number of blocks in the compressed address, subtract this number from 8, and then multiply the result by 16. For example, in the address FF02::2, there are two blocks (the FF02 block and the 2 block); the number of bits expressed by the :: is 96 (= (8 – 2) × 16) [MSD200401].

2.2.2 Addressing Issues/Reachability

Every IPv6 address has a defined reachability scope. Table 2.1 shows the address and associated reachability scopes. The reachability of *node-local addresses* is "the same node"; the reachability of *link-local addresses* is "the local link"; the reachability of *site-local addresses*[1] is "the private intranet"; and the reachability of *global addresses* is "the IPv6-enabled Internet." IPv6 interfaces can have multiple addresses that have different reachability scopes. For example, a node may have a link-local address, a site-local address, and a global address.

Similar to the IPv4 address space, the IPv6 address space is partitioned according to the value of high-order bits (known as a format prefix) in the address. Table 2.2 depicts the IPv6 address space allocation by format prefixes. The (current) set of unicast addresses that can be employed by IPv6 nodes consists of aggregatable global unicast addresses, link-local unicast addresses, and site-local unicast addresses (these addresses represent about 12.6 percent of the entire IPv6 address space, but it is still ~ 3.4×10^{37}). The prefix is the portion of the address that indicates the bits that have fixed values or are the bits of the network identifier. Prefixes

---

[1] Site-local has been deprecated by RFC3879 but is addressed herewith for historical purposes.

Table 2.1 IPv6 Address and Associated Reachability Scopes

Address Scope/ Reachability	Description
Node-local addresses to reach same node	Used to send protocol data units (PDUs) to the same node: • Loopback address (PDUs addressed to the loopback address are never sent on a link or forwarded by an IPv6 router — this is equivalent to the IPv4 loopback address) • Node-local multicast address
Link-local addresses to reach local link[a]	Used to communicate between host devices (e.g., servers, VoIP devices, etc.) on the link; these addresses are always configured automatically: • Unspecified address. It indicates the absence of an address and is typically used as a source address for PDUs that are attempting to verify the uniqueness of a tentative address (it is equivalent to the IPv4 unspecified address). The unspecified address is never assigned to an interface or used as a destination address. • Link-local unicast address • Link-local multicast address
Site-local addresses to reach the private intranet (internetwork)[a]	Used between nodes that communicate with other nodes in the same site; site-local addresses are configured by router advertisement: • Site-local Unicast address — these addresses are not reachable from other sites, and routers must not forward site-local traffic outside of the site. Site-local addresses can be used in addition to aggregatable global unicast addresses. • Site-local Multicast address
Global addresses to reach the Internet (IPv6 enabled); also known as aggregatable global unicast addresses	Globally routable and reachable addresses on the IPv6 portion of the Internet (they are equivalent to public IPv4 addresses); global addresses are configured by router advertisement: • Global Unicast address • Other scope Multicast address Global addresses are designed to be aggregated or summarized to produce an efficient, hierarchical addressing and routing structure.

[a] When one specifies a link-local or site-local address, one needs to also specify a scope ID, which further defines the reachability scope for these (nonglobal) addresses.

Table 2.2   IPv6 Address Space Allocation

Address Space Allocation	Format Prefix	Percentage of the Address Space	Hex Notation	Fraction of the Address Space
Reserved	0000 0000	0.391%	0x00	1/256
Reserved for NSAP allocation	0000 001	0.781%	0x0 001	1/128
Aggregatable global unicast addresses	001	12.500%	001	1/8
Link-local unicast addresses	1111 1110 10	0.098%	0xFE 10	1/1024
Site-local unicast addresses	1111 1110 11	0.098%	0xFE 11	1/1024
Multicast addresses	1111 1111	0.391%	0xFF	1/256
The remainder of the IPv6 address	Unassigned	85.742%		

Note: 0xY is the hexadecimal notation for digit "Y."

for IPv6 routes and subnet identifiers are expressed in the same way as classless interdomain routing notation for IPv4. An IPv6 prefix is written in address/prefix-length notation. (IPv4 environments use a dotted-decimal representation known as the subnet mask to establish the network prefix of a given IP address; the subnet mask approach is *not used* in IPv6; rather, only the prefix-length notation is used.)

As noted, the prefix is the part of the address that indicates the bits that have fixed values or are the bits of the network identifier. For example:

21DA:D3::/48 is a 48-bit route prefix

21DA	00D3	0000	16 bits	16 bits	16 bits	16 bits	16 bits
<- route prefix ->							

and

21DA:D3:0:2F3B::/64 is a 64-bit subnet prefix

(a 48-bit route prefix plus a site topology identifier for the next 16 bits)

21DA	00D3	0000	2F3B	16 bits	16 bits	16 bits	16 bits
<- route prefix ->			<- subnet prefix ->	16 bits	16 bits	16 bits	16 bits

Table 2.3 Address Scope versus Reachability Scope

Address Scope	Reachability Scope
Node local	Same node
Link local	Local link (LAN)
Site local	A private internetwork (intranet)

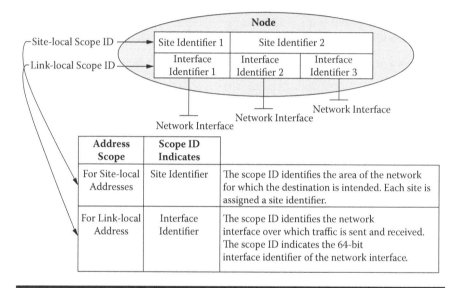

Figure 2.1 Example of logical node in IPv6.

2.2.3 Scope/Reachability

The scope ID identifies a specific area within the reachability scope for nonglobal addresses (recall that the reachability scope is related to the address scope as shown in Table 2.3). A node identifies each area of the same scope with a unique scope ID.

Figure 2.1 shows an example of how the scope ID indicates an interface or site identifier, depending on the scope of the address. In this example, the node is connected to three links and two sites. Here,

■ The sites (specifically, site identifiers 1 and 2) are identified by the site-local (intranet) scope ID.
■ The links (specifically, interface identifiers 1, 2, and 3) are identified by the link-local (local area network, LAN) scope ID.

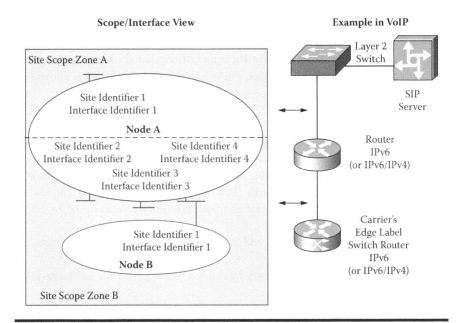

Figure 2.2   How the nodes use the scope ID to identify site scope zones.

The notation utilized by some to specify the scope ID with an address is *Address%ScopeID*. Figure 2.2 depicts an example of how the nodes use the scope ID to identify site scope zones. As one can see in the figure, the interface identifier scope ID is used only by the local node; other nodes may use a different network interface or site identifier for the same link, for example, the link that has scope ID (interface identifier) 4 for Node A has scope ID 1 for Node B. For the example of Figure 2.2, the following describes the link-local address FE70::3 qualified with a scope ID on the link between Node A and Node B:

- For Node A, the address is FE70::3%4.
- For Node B, the address is FE70::3%1.

Each attached zone of the same scope must be assigned a different site identifier, but attached zones of different scopes can reuse the same index.

Implicit in Figure 2.2 is a topology hierarchy, as follows:

- Public topology: The collection of larger and smaller ISPs that provide access to the IPv6 Internet.
- Site topology: The collection of subnets within an organization's site (namely, this is the intranet, although it need not be strictly contained at a single location — the term *site* here has more an implication of an organization's domain than a single physical site).
- Interface identifier topology: Identifies a specific interface on a subnet within an organization's site.

2.3 Address Types

This section looks at some more detailed information related to address types. We discuss a number of unicast addresses, multicast addresses, and anycast addresses.

2.3.1 Unicast IPv6 Addresses

A unicast address identifies a single interface within the scope of the unicast address type. This could be a Voice-over-IP (VoIP) handset in a VoIPv6 environment, a personal computer (PC) on a LAN, and so on. Utilizing an up-to-date unicast routing topology, protocol data units (PDUs) addressed to a unicast address are delivered to a single interface. Unicast addresses fall into the following categories:

■ Aggregatable global unicast addresses (e.g., used to reach an Internet-connected VoIP phone)
■ Link-local addresses (e.g., used to reach a VoIP phone on the same LAN segment)
■ Site-local addresses (e.g., used to reach a VoIP phone on a corporate intranet)
■ Special addresses, including unspecified and loopback addresses
■ Compatibility addresses, including 6to4 addresses

These are discussed next.

2.3.1.1 Aggregatable Global Unicast Addresses

The IPv6-based Internet has been designed to support efficient, hierarchical addressing and routing (this is in contrast to IPv4-based Internet, which has a mixture of both flat and hierarchical routing). Aggregatable global unicast addresses are globally routable and globally reachable on the IPv6 portion of the (IPv6) Internet. The region of the Internet over which the aggregatable global unicast address is unique (the scope) is the entire IPv6 Internet. As we saw in Table 2.2, aggregatable global unicast addresses (also known as global addresses) are identified by the format prefix of 001. This type of addressing can be used, for example, to reach an Internet-connected VoIP (SIP, Session Initiation Protocol) phone (say, the author's phone given to him by his company and utilized by him while traveling on business and using the Internet for connectivity) from any origination point, whether such origination point is on the firm's intranet, on any other company's intranet, or even at another Internet point. This enables the end-to-end connectivity that we have alluded to earlier in this book.

Figure 2.3 shows how the fields within the aggregatable global unicast address create a three-level topological structure with globally unique addresses. The first 48 bits are comprised of the 3-bit format prefix; the top-level aggregator (TLA)[2]

---

[2] TLA/NLA/SLA field structure has been deprecated by RFC3587 but is covered herewith for historical purposes.

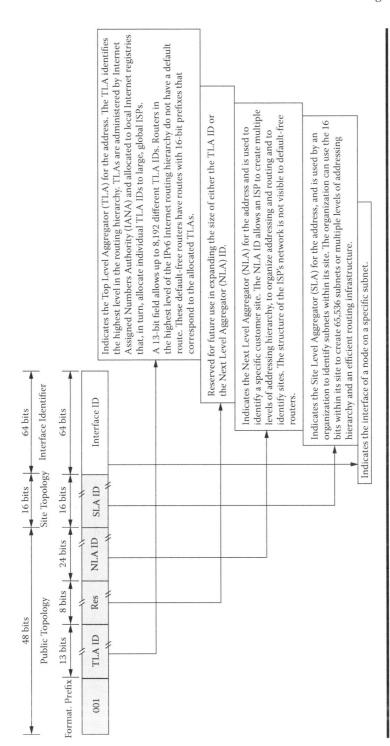

Figure 2.3   Aggregatable global unicast address.

ID comprises the next 13 bits; the next 8 bits are reserved; and the next 24 bits represent the next-level aggregator (NLA) ID. This combination gives the first two levels. The next 16 bits represent the site topology, namely, the site-level aggregator (SLA) ID. The SLA is used by a firm or organization to identify subnets within its site (intranet); the organization can use the 16 bits within its site to create 65,536 subnets or multiple levels of addressing hierarchy, which can also facilitate the routing process. (Note that with a two-octet address space for subnetting, an aggregatable global unicast prefix assigned to a firm is equivalent to granting that firm an IPv4 Class A network ID. Also, the structure of the customer's network is not visible to the Internet Service Provider (ISP).) Finally, the interface ID points to the interface of a node on a specific subnet.

Addresses of this type can, by design, be aggregated (summarized) to produce an efficient routing infrastructure.

2.3.1.2 Link-Local (Unicast) Addresses

Link-local addresses are utilized by nodes when communicating with neighboring nodes on the same link. For example, link-local addresses are used to communicate between hosts on the link on a single-link IPv6 network without the intervention/utilization of a router (e.g., in a LAN segment, a virtual LAN (VLAN), etc.). This type of addressing can be used to reach a company colleague on a LAN-connected VoIP phone (say, for colleagues working in the same building — assuming that both are on the same LAN).

The scope of a link-local address is the local link. An IPv6 router does not forward link-local traffic beyond the link. A link-local address is required for Neighbor Discovery (ND) processes and is always automatically configured, even in the absence of all other unicast addresses. As seen in Table 2.2, link-local addresses are identified by the format prefix of 1111 1110 10. The address starts with FE (for example, 1111 1110 1000 is 0xFE8; 1111 1110 1001 is 0xFE9; 1111 1110 1010 is 0xFEA; and 1111 1110 1011 is 0xFEB.) With the 64-bit interface identifier, the prefix for link-local addresses is, by convention, always FE80::/64.

2.3.1.3 Site-Local (Unicast) Addresses

Site-local addresses are utilized between nodes that communicate with other nodes in the same site (organization). The scope of a site-local address is the site, which is the organization intranet (internetwork). This type of addressing can be used to reach a company colleague on an intranet-connected VoIP phone, say for colleagues working in the same company but perhaps at two company locations in two cities (in this arrangement, however, those VoIP phones would not be directly reachable from anywhere that happens to be on an IP network [see next paragraph]; they may, nonetheless, be reachable through a gateway).

As seen in Table 2.2, site-local addresses are identified by the format prefix of 1111 1110 11 (they are equivalent to the IPv4 private address space 10.0.0.0/8, 172.16.0.0/12, and 192.168.0.0/16). Hence, if there are private intranets that do not have a direct, routed connection to the IPv6 Internet, they can use site-local addresses without conflicting with aggregatable global unicast addresses. However, one should keep in mind that increasingly virtually all private intranets have connections to the Internet. Also, if end-to-end anytime, anyplace, VoIP communication is to be supported, aggregatable global unicast may be indicated for each VoIP device. This is because site-local addresses are not reachable from other sites (namely, other organizations), and routers must not forward site-local traffic outside the site (organization). Fortunately, aggregatable global unicast addresses may be assigned to VoIP devices in addition to site-local addresses. Unlike link-local addresses, site-local addresses are not automatically configured and must be assigned through the stateless address configuration process [MSD200401].

Referring to Figure 2.1, one notes that the first 48 bits are always fixed for site-local addresses, beginning with FEC0::/48. Beyond the 48 fixed bits is a 16-bit subnet identifier (subnet ID field) with which the network administrator can create subnets for use within the organization (one can create up to 65,536 subnets in a flat subnet structure, or one can partition the high-order bits of the subnet ID field to create a hierarchical and aggregatable routing infrastructure). Beyond the subnet ID field is a 64-bit interface ID field that identifies a specific interface on a subnet.

Notice from this discussion that the aggregatable global unicast address and the site-local address share the same structure beyond the first 48 bits of the address. In aggregatable global unicast addresses, the SLA ID identifies the subnet within an organization, and for site-local addresses, the subnet ID performs the same function. Because of this characteristic, one can assign a specific subnet number to identify a subnet that is used for both site-local and aggregatable global unicast addresses.

2.3.1.4 Unspecified (Unicast) Address

The unspecified address 0:0:0:0:0:0:0:0 (that is, ::) indicates the absence of an address and is typically used as a source address for PDUs that are attempting to verify the uniqueness of a tentative address. It is equivalent to the IPv4 unspecified address of 0.0.0.0. The unspecified address is never assigned to an interface or used as a destination address.

2.3.1.5 Loopback (Unicast) Address

The loopback address 0:0:0:0:0:0:0:1 or ::1 identifies a loopback interface, enabling a node to send PDUs to itself. It is equivalent to the IPv4 loopback address of 127.0.0.1. PDUs addressed to the loopback address are never sent on a link or forwarded by an IPv6 router.

2.3.1.6 Compatibility (Unicast) Addresses

IPv6 provides what are called *6to4 addresses* to facilitate the coexistence of IPv4-to-IPv6 environments and the migration from the IPv4 to the IPv6 environment. The 6to4 address is used for communicating between two nodes operating both IPv4 stacks and IPv6 stacks over an IPv4 routing infrastructure (more on this in Chapter 6). The 6to4 address is formed by combining the prefix 2002::/16 with the 32 bits of the public IPv4 address of the node, forming a 48-bit prefix.

2.3.2 Multicast IPv6 Addresses

A useful feature supported in IPv6 is multicasting. The use of multicasting in IP networks is defined in RFC 1112, which describes addresses and host extensions for the way IP hosts support multicasting — the concepts originally developed for IPv4 also apply to IPv6. Besides a variety of protocol-level functionality supported by multicasting (e.g., Multicast Listener Discovery [MLD] and ND), one also can use this mechanism to support VoIP/IPTV (IP-based TV) functionality (e.g., audio-conferencing/bridging and program distribution). Multicast traffic is promulgated by utilizing a single destination address in the IPv6 header but the IPv6 datagram is received and processed by multiple hosts. Hosts and devices listening on a specific multicast address comprise a multicast group; these devices receive and process traffic sent to the group address. As seen in Table 2.2, IPv6 multicast addresses have the format prefix of 1111 1111; namely, the multicast address always begins with 0xFF.

Group membership in multicast is dynamic, allowing hosts to join and leave the group at any time. Groups can be from multiple network segments (links or subnets) if the connecting routers support forwarding of multicast traffic and group membership information [MSD200401]. A host (e.g., a VoIP SIP proxy or an H.323 gatekeeper) can send traffic to a group address without belonging to the group. In fact, to join a group, a host sends a group membership message. Each multicast group is identified by one IPv6 multicast address. All group members who listen and receive IPv6 messages sent to the group address share the group address. Multicast routers periodically poll membership status.

Some of the reserved IPv6 multicast addresses (RFC 2375) are shown in Table 2.4.

A multicast address is an addressing mechanism that identifies multiple interfaces; it is used for one-to-many communication. With the appropriate multicast routing topology, PDUs addressed to a multicast address are delivered to all interfaces that are identified by the address. Multicast addresses cannot be utilized as source addresses. Multicast address Flags, Scope, and Group are shown in Figure 2.4.

To identify all nodes for the node-local and link-local scopes, the following multicast addresses are defined:

Table 2.4   Reserved Multicast IPv6 Addresses

IPv6 Multicast Address	Description
FF02::1	The all-nodes address used to reach all nodes on the same link.
FF02::2	The all-routers address used to reach all routers on the same link.
FF02::4	The address used to reach all Distance Vector Multicast Routing Protocol (DVMRP) multicast routers on the same link.
FF02::5	The address used to reach all Open Shortest Path First (OSPF) routers on the same link.
FF02::1:FF*XX:XXXX*	The solicited-node address used in the address resolution process to resolve the IPv6 address of a link-local node to its link-layer address. The rightmost 24 bits (*XX:XXXX*) of the solicited-node address are the rightmost 24 bits of an IPv6 unicast address.

- FF01::1 (node-local scope all-nodes address)
- FF02::1 (link-local scope all-nodes address)

To identify all routers for the node-local, link-local, and site-local scopes, the following multicast addresses are defined:

- FF01::2 (node-local scope all-routers address)
- FF02::2 (link-local scope all-routers address)
- FF05::2 (site-local scope all-routers address)

Next, we briefly look at solicited-node addresses. The solicited-node address supports efficient querying of network nodes for the purpose of address resolution. IPv6 uses the Neighbor Solicitation message to perform address resolution. This multicast address consists of the prefix FF02::1:FF00:0/104 along with the last 24 bits of the IPv6 address that is being resolved. In contrast to IPv4, by which the ARP request frame is sent via a broadcast at the Media Access Control (MAC) level, and in doing so imposing on all nodes on the network segment, in IPv6 the solicited-node multicast address is used as the Neighbor Solicitation message destination. This avoids imposing on all IPv6 nodes on the local link by using the local-link scope all-nodes address.

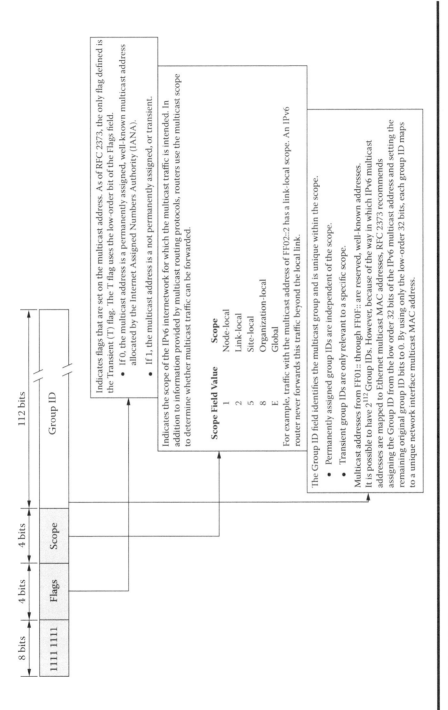

Figure 2.4 Multicast address.

2.3.3 Anycast IPv6 Addresses

An anycast address identifies multiple interfaces but not an entire broadcast universe. This could be used, for example, to support VoIP voice mail group distribution. With the appropriate routing topology, PDUs addressed to an anycast address are delivered to a single interface for further appropriate handling (a PDU addressed to an anycast address is delivered to the nearest interface identified by the address). To make possible the delivery to the nearest anycast group member, the routing infrastructure must be aware of the interfaces that are assigned anycast addresses and must know their distances in terms of routing metrics. At present, anycast addresses are used only as destination addresses and are assigned only to routers.

2.4 Addresses for Hosts and Routers

In contrast to IPv4, by which a host with a single network adapter has a single IPv4 address assigned to that adapter, an IPv6 host (e.g., a SIP proxy) typically has multiple IPv6 addresses (even in the case of a single interface). (When a computer is configured with more than one IP address, it is referred to as a *multihomed* system.) IPv6 host and router address usage is as follows [MSD200401]:

Host: Typical IPv6 hosts are logically multihomed because they have at least two addresses with which they can receive PDUs. Each host is assigned the following unicast addresses:

- A link-local address for each interface. This address is used for local traffic.
- An address for each interface. This could be a routable site-local address and one or more global addresses.
- The loopback address (::1) for the loopback interface.

In addition, each host is listening for traffic on the following multicast addresses:

- The node-local scope all-nodes address (FF01::1).
- The link-local scope all-nodes address (FF02::1).
- The solicited-node address for each unicast address on each interface.
- The multicast addresses of joined groups on each interface.

Router: An IPv6 router is assigned the following unicast addresses:

- A link-local address for each interface. This address is used for local traffic.
- An address for each interface. This could be a routable site-local address and one or more global addresses.
- The loopback address (::1) for the loopback interface.

An IPv6 router is assigned the following anycast addresses:

- A subnet-router anycast address for each subnet.
- Additional anycast addresses (optional).

Each router is listening for traffic on the following multicast addresses:

- The node-local scope all-nodes address (FF01::1).
- The node-local scope all-routers address (FF01::2).
- The link-local scope all-nodes address (FF02::1).
- The link-local scope all-routers address (FF02::2).
- The site-local scope all-routers address (FF05::2).
- The solicited-node address for each unicast address on each interface.
- The addresses of joined groups on each interface.

2.4.1 Interface Determination

As noticed in Figure 2.1, the last 64 bits of an IPv6 address are the interface identifier that is unique to the 64-bit prefix of the IPv6 address. There are two ways for interface identifier determination: (1) derivation from the Institute of Electrical and Electronics Engineers (IEEE) extended unique identifier (EUI)-64 address; and (2) by random generation and random change over time. IETF RFC 2373 stipulates that unicast addresses that use format prefixes 001 through 111 must use a 64-bit interface identifier that is derived from the EUI-64 address. Related to the second approach, RFC 3041 states that, to provide a level of anonymity, the identifier can be randomly generated and changed over time.

EUI-64 addresses are either assigned to a network adapter or derived from IEEE 802 addresses. LAN network interface cards (NICs) that (at this point in the development of hardware) typically comprise the physical interface (network adapters) of host and device identifiers use the 48-bit IEEE 802 address. This address (also called the physical, hardware, or MAC address) consists of two parts: company ID and extension ID. The company ID is 24-bit ID uniquely assigned to each manufacturer of network adapters; this is also known as the manufacturer ID. The extension ID (also known as the board ID) is a 24-bit ID uniquely assigned to each network adapter at the time of assembly. The IEEE 802 address is thus a globally unique 48-bit address. The IEEE EUI-64 address is a newly defined standard for network interface addressing. The company ID is 24 bits in length, but the extension ID is 40 bits, supporting a larger address space for a network adapter manufacturer (see Figure 2.5).

Note that, for a typical 802.x network adapter address, both U/L and I/G bits are set to 0, corresponding to a universally administered, unicast MAC address.

To generate an EUI-64 address from an IEEE 802 address, 16 bits of 11111111 11111110 (0xFFFE) are inserted into the IEEE 802 address between the company ID and the extension ID (see Figure 2.6).

2.4.2 Mapping EUI-64 Addresses to IPv6 Interface Identifiers

An IPv6 unicast address utilizes a 64-bit interface identifier. To obtain this identifier from a EUI-64 address, the U/L bit in the EUI-64 address is complemented

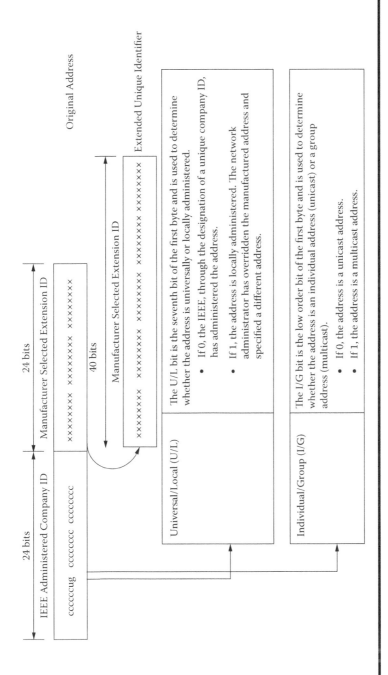

Figure 2.5 IEEE address along with the extended unique identifier.

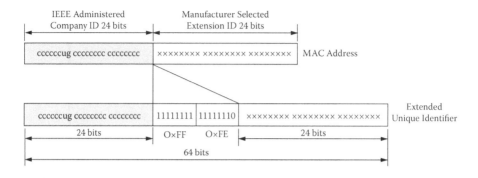

Figure 2.6   Extended unique identifier generated from MAC address.

(if it is a 1, it is set to 0; if it is a 0, it is set to 1). The resulting bitstream is used as a universally administered unicast EUI-64 address.

2.4.3 Mapping IEEE 802 Addresses to IPv6 Interface Identifiers

To obtain an IPv6 interface identifier from an IEEE 802 address, one must first map the IEEE 802 address to an EUI-64 address, as discussed previously; then, one must complement (flip) the U/L bit. The resulting bitstream is used as a universally administered unicast IEEE 802 address.

2.4.4 Randomly Generated Interface Identifiers

IPv6 interface identifiers remain static over time; hence, for security reasons, a capability is needed to generate temporary addresses. (Because of network address translation/Dynamic Host Control Protocol (NAT/DHCP), in an IPv4 environment it is difficult to track a user's traffic on the basis of IP address.) However, it should be noted that many — if not most — hacking techniques do not rely on knowing the IP address of a *specific* device on a network; instead, such techniques simply look for any available entry point. After that, a deposited Trojan horse may do the job of perpetrating a full infraction.

In IPv6 after the connection is made through router discovery and stateless address autoconfiguration, the end-user device is assigned a 64-bit prefix. If the interface identifier is based on a EUI-64 address (which, as we discussed, is derived from the static IEEE 802 address), the traffic of a specific node can be identified, which opens up the possibility of tracking a specific user should that be of interest to an intruder. To address this issue, an alternative IPv6 interface identifier can be randomly generated and changed over time, as described in [RFC3041]. For IPv6 systems that have storage capabilities, a history value is stored. When IPv6 is initialized, a new interface identifier is created through the following process (the IPv6 address based on this random interface identifier is known as a temporary address):

1. Retrieve the history value from storage and append the interface identifier based on the EUI-64 address of the adapter.
2. Compute the message digest 5 (MD5) one-way encryption hash over the quantity in Step 1.
3. Save the last 64 bits of the MD5 hash computed in Step 2 as the history value for the next interface identifier computation.
4. Take the first 64 bits of the MD5 hash computed in Step 2 and set the seventh bit to zero. The seventh bit corresponds to the U/L bit that, when set to 0, indicates a locally administered interface identifier. The result is the interface identifier.

Temporary addresses are generated for public address prefixes that use stateless address autoconfiguration.

References

[RFC3041] T. Narten, R. Daves, Private Extensions for Stateless Address Autoconfiguration in IPv6, RFC 3041, January 2001.

[RFC3315] R. Drome, J. Bound, B. Volz, T. Lemon, C. Perkins, M. Carney, Dynamic Host Configuration Protocol for IPv6 (DHCPv6), RFC 3315, July 2003.

[RFC4291] R. Hinden and S. Deering, Internet Protocol Version 6 (IPv6) Addressing Architecture, RFC 4291, February 2006.

[MSD200401] Microsoft Corporation, MSDN Library, Internet Protocol, 2004, http://msdn.microsoft.com.

Chapter 3

IPv6 Network Constructs

3.1 Introduction

The Internet Protocol version 6 (IPv6) Specification [RFC2460] and the IPv6 Addressing Architecture [RFC2373] provide the base architecture and design of IPv6; we covered some of these key concepts in Chapters 1 and 2. This chapter looks at basic IPv6 network constructs, specifically routing processes. Because there are differences in some of the details of how these IPv6 processes operate compared with IPv4, it is worth looking at some of these issues. This chapter covers logical networking issues; other chapters in the book focus more directly on actual enterprise/institutional networking-level issues. Related work in IPv6 that needs to be mastered by implementors and network designers (covered in chapters that follow) includes the IPv6 stateless address autoconfiguration [RFC2462]; IPv6 Neighbor Discovery (ND) processing [RFC2461]; the Dynamic Host Control Protocol for IPv6 (DHCPv6) [RFC3315]; and the Dynamic Updates to DNS (Domain Name System) [RFC2136]. Some portions of this discussion are based on [MSD200401].

3.2 IPv6 Infrastructure

3.2.1 Protocol Mechanisms

As discussed in Chapter 1, an IPv6 protocol data unit (PDU) consists of an IPv6 header and an IPv6 payload, as depicted in Figure 3.1. The IPv6 header consists

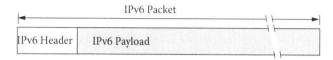

IPv6 Header Field	Function	Length
Version	Identifies the version of the protocol. For IPv6, the field is 6 (1010)	4 bits
Class	Intended for originating nodes and forwarding routers to identify and distinguish between different classes or priorities of IPv6 packets.	8 bits
Flow Label	Defines how traffic is handled and identified. A flow is a sequence of packets sent either to a unicast or a multicast destination. Field identifies packets that require special handling by the IPv6 node.	20 bits
Payload Length	Identifies the length, octets, of the payload. This field is a 16-bit unsigned integer. The payload includes the optional extension headers, as well as the upper-layer protocols.	16 bits
Next Header	Identifies the header immediately following the IPv6 header.	8 bits
Hop Limit	Identifies the number of network segments (links or subnets), on which the packet is allowed to travel before being discarded by a router. This parameter is set by the sending host and is used to prevent packets from endlessly circulating on an IPv6 internetwork. When forwarding an IPv6 packet, IPv6 routers must decrease the Hop Limit by 1, and must discard the IPv6 packet when the Hop Limit is 0.	8 bits
Source Address	Identifies the IPv6 address of the original source of the IPv6 packet.	128 bits
Destination Address	Identifies the IPv6 address of the intermediate or final destination of the IPv6 packet.	128 bits

Figure 3.1   IPv6 protocol data unit (PDU).

of two parts: the IPv6 base header and optional extension headers. The optional extension headers are considered part of the IPv6 payload, as are the Transmission Control Protocol/User Datagram Protocol/Real-Time Transport Protocol (TCP/UDP/RTP) PDUs. Obviously, IPv4 headers and IPv6 headers are not automatically interoperable; hence, a router operating in a mixed environment must support an implementation of both IPv4 and IPv6 to deal with both header formats. Figure 3.2 shows for illustration purposes the flows of IPv6 PDUs in a Voice-over-IP (VoIP) environment.

As we noted in Chapter 2, the large size of the IPv6 address allows it to be subdivided into hierarchical routing domains that are supportive of the topology of today's ubiquitous Internet (IPv4-based Internet lacks this flexibility). Conveniently, the use of 128 bits provides multiple levels of hierarchy and flexibility in designing hierarchical addressing and routing.

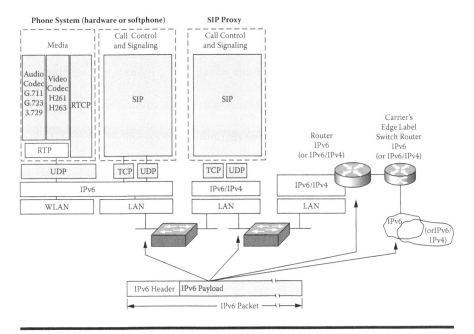

Figure 3.2   Flows of IPv6 packets in a VoIPv6 environment.

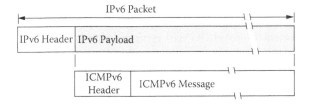

Figure 3.3   An ICMPv6 message.

3.2.2 Protocol Support Mechanisms

Two support mechanisms are of interest: a mechanism to deal with communication transmission issues and a mechanism to support multicast.

Internet Control Messaging Protocol for IPv6 (ICMPv6) (defined in [RFC2463]) is designed to enable hosts and routers that use IPv6 protocols to report errors and forward other basic status messages. For example, ICMPv6 messages are sent by network elements when an IPv6 PDU cannot be forwarded further along to reach its intended destination. ICMPv6 messages are carried as the payload of IPv6 PDUs (see Figure 3.3); hence, there is no guarantee of their delivery.

The following list identifies the functionality supported by the basic ICMPv6 mechanisms:

- Destination unreachable: An error message that informs the sending host that a PDU cannot be delivered.
- Packet too big: An error message that informs the sending host that the PDU is too large to forward.
- Time exceeded: An error message that informs the sending host that the hop limit of an IPv6 PDU has expired.
- Parameter problem: An error message that informs the sending host that an error was encountered in processing the IPv6 header or an IPv6 extension header.
- Echo request: An informational message that is used to determine whether an IPv6 node is available on the network.
- Echo reply: An informational message that is used to reply to the ICMPv6 echo request message.

The ping command is basically an ICMPv6 echo request message along with the receipt of ICMPv6 echo reply messages. Just as is the case with IPv4, one can use pings to detect network or host communication failures and troubleshoot connectivity problems.

ICMPv6 also supports Multicast Listener Discovery (MLD). MLD ([RFC2710], [RFC3590], and [RFC3810]) enables one to manage subnet multicast membership for IPv6. MLD is a collection of three ICMPv6 messages that replace the Internet Group Management Protocol (IGMP) version 3 that is employed in IPv4. MLD messages are used to determine group membership on a network segment, also known as a link or subnet. As implied, MLD messages are sent as ICMPv6 messages. They are used in the context of multicast communications (see below):

- Multicast listener query: Message issued by a multicast router to poll a network segment for group members. Queries can be general, requesting group membership for all groups, or can request group membership for a specific group.
- Multicast listener report: Message issued by a host when it joins a multicast group or in response to an MLD multicast listener query sent by a router.
- Multicast listener done: Message issued by a host when it leaves a host group and is the last member of that group on the network segment.

ICMPv6 also supports ND. ND [RFC2461] is a collection of five ICMPv6 messages that manage node-to-node communication on a link. Nodes on the same link are also called neighboring nodes. ND replaces Address Resolution Protocol (ARP), ICMPv4 Router Discovery, and the ICMPv4 redirect message. Table 3.1 identifies key ND processes [MSD200401]. Hosts (e.g., servers, Session Initiation

Table 3.1   Key ND Processes

Process	Description
Address autoconfiguration	The process for configuring IP addresses for interfaces in the absence of a stateful address configuration server, such as via Dynamic Host Control Protocol version 6 (DHCPv6).
Address resolution	The process by which a node resolves a neighboring node's IPv6 address to its link-layer address. The resolved link-layer address becomes an entry in a neighbor cache in the node. The link-layer address is equivalent to ARP in IPv4, and the neighbor cache is equivalent to the ARP cache. The neighbor cache displays the interface identifier for the neighbor cache entry, the neighboring node IPv6 address, the corresponding link-layer address, and the state of the neighbor cache entry.
Duplicate address detection	The process by which a node determines that an address considered for use is not already in use by a neighboring node. This is equivalent to the use of ARP frames in IPv4.
Dynamic Updates to DNS [RFC2136]	A process that supports the dynamic update of DNS records for both IPv4 and IPv6. DHCP can use the dynamic updates to DNS to integrate addresses and name space to support not only autoconfiguration but also autoregistration in IPv6 [RFC3315].
IPv6 Neighbor Discovery [RFC2461]	The node discovery process/protocol in IPv6 that replaces and enhances functions of ARP. To understand IPv6 and stateless address autoconfiguration, implementors and network designers need to understand IPv6 Neighbor Discovery [RFC3315].
Neighbor unreachability detection	The process by which a node determines that neighboring hosts or routers are no longer available on the local network segment. After the link-layer address for a neighbor has been determined, the state of the entry in the neighbor cache is tracked. If the neighbor is no longer receiving and sending back PDUs, the neighbor cache entry is eventually removed.
Next-hop determination	The process by which a node determines the IPv6 address of the neighbor to which a PDU is being forwarded. The determination is made based on the destination address. The forwarding or next-hop address is either the destination address of the PDU being sent or the address of a neighboring router.

Table 3.1   Key ND Processes (continued)

Process	Description
Next-hop determination (continued)	The resolved next-hop address for a destination becomes an entry in a node's destination cache, also known as a route cache. The route cache displays the destination address, the interface identifier and next-hop address, the interface identifier and address used as a source address when sending to the destination, and the path maximum transfer unit (MTU) for the destination.
Parameter discovery	The process by which a host discovers additional operating parameters, including the link MTU and the default hop limit for outbound PDUs.
Prefix discovery	The process by which a host discovers the network prefixes for local destinations.
Redirect function	The process by which a router informs a host of a better first-hop IPv6 address to reach a destination. This is equivalent to the function of the IPv4 ICMP redirect message.
Router discovery	The process by which a host discovers the local routers on an attached link and automatically configures a default router. In IPv4, this is equivalent to using ICMPv4 router discovery to configure a default gateway.

Protocol [SIP] proxies, H.323 gatekeepers, etc.) make use of ND to discover neighboring routers, addresses, address prefixes, and other configuration parameters. Routers make use of ND to advertise their presence, host configuration parameters, and on-link prefixes. Routers also use ND to inform hosts of a better next-hop address to forward PDUs for a specific destination. Nodes make use of ND to resolve the link-layer address of a neighboring node to which an IPv6 PDU is being forwarded. Nodes also use ND to determine when the link-layer address of a neighboring node has changed and whether IPv6 PDUs can be sent to and received from a neighbor.

3.3 Routing and Route Management

Routing is the process of forwarding PDUs between connected network segments (also known as links or subnets). Routing is a primary function of a network-layer protocol, whether it is IP version 4 or version 6. IPv6 routers provide the primary means for joining together two or more IPv6 network segments. Network segments are identified by using an IPv6 network prefix and prefix length. Routers pass IPv6 PDUs from one network segment to another. IPv6 routers are attached to two or more IPv6 network segments and enable hosts on those segments to forward IPv6 PDUs. IPv6 PDUs are exchanged and processed on each host by using IPv6 at Layer 3 (the Internet Protocol layer).

Datagrams with a source and destination IP address identified in the header are handed to the IP engine/layer. Above the IPv6 layer, transport services on the source host pass data in the form of TCP segments or UDP PDUs down to the IPv6 layer. IPv6 layer services on each sending host examine the destination address of each PDU, compare this address to a locally maintained routing table, and then determine what additional forwarding is required. The IPv6 layer creates IPv6 PDUs with source and destination address information that is used to route the data through the network. The IPv6 layer then passes PDUs down to the link layer, where the PDUs are converted into frames for transmission over network-specific media on a physical network. This process occurs in reverse order on the destination host [MSD200401].

IPv6 hosts utilize routing tables to maintain information about other IPv6 networks and IPv6 hosts. The routing tables provide important information about how to communicate with remote networks and hosts. Every device that implements IPv6 determines how to forward PDUs based on the contents of the IPv6 routing table. The following information is contained in the IPv6 routing table:

- An address prefix
- The interface over which PDUs that match the address prefix are sent
- A forwarding or next-hop address
- A preference value used to select between multiple routes with the same prefix
- The lifetime of the route
- The specification of whether the route is published (advertised in a routing advertisement)
- The specification of how the route is aged
- The route type

The IPv6 routing table is built automatically, based on the current IPv6 configuration of the router. When forwarding IPv6 PDUs, the router searches the routing table for an entry that is the most specific match to the destination IPv6 address. A route for the link-local prefix (FE80::/64) is not displayed.

Typically, a default route is used by an end device because it is not practical for an end device to maintain a routing table for each communication device on an IPv6 network. The default route (a route with a prefix of ::/0) is typically used to forward an IPv6 PDU to a default router on the local link. Because the router that corresponds to the default router contains information about the network prefixes of the other IPv6 subnets within the larger IPv6 internetwork, it forwards the PDU to other routers until the PDU is eventually delivered to the destination.

The following steps occur during the routing process [MSD200401]:

1. Before a communication device sends an IPv6 PDU, it inserts its source IPv6 address and the destination IPv6 address (for the recipient) into the IPv6 header.
2. The device then examines the destination IPv6 address, compares it to a locally maintained IPv6 routing table, and takes appropriate action. The device does one of the following:
   ■ It passes the PDU to a protocol layer above IPv6 on the local host.
   ■ It forwards the PDU through one of its attached network interfaces.
   ■ It discards the PDU.
3. IPv6 searches the routing table for the route that is the closest match to the destination IPv6 address. The most specific to the least-specific route is determined in the following order:
   ■ A route that matches the destination IPv6 address (a host route with a 128-bit prefix length).
   ■ A route that matches the destination with the longest prefix length.
   ■ The default route (the network prefix ::/0).
4. If a matching route is not found, the destination is determined to be an on-link destination.

References

[MSD200401] Microsoft Corporation, MSDN Library, Internet Protocol, 2004, http:// msdn.microsoft.com.
[RFC2136] P. Vixie, S. Thomson, Y. Rekhter, J. Bound, Dynamic Updates in the Domain Name System (DNS UPDATE), RFC 2136, April 1997.
[RFC2373] R. Hinden, S. Deering, IP Version 6 Addressing Architecture, RFC 2373, July 1998.
[RFC2460] S. Deering, R. Hinden, Internet Protocol, Version 6 (IPv6) Specification, RFC 2460, December 1998.
[RFC2461] T. Narten, E. Nordmark, W. Simpson, Neighbor Discovery for IP Version 6 (IPv6), RFC 2461, December 1998.
[RFC2462] S. Thompson, T. Narten, IPv6 Stateless Address Autoconfiguration, RFC 2462, December 1998.
[RFC2463] A. Conta, S. Deering, Internet Control Message Protocol Version 6 (ICMPv6) for the Internet Protocol Version 6(IPv6) Specification, RFC 2463, December 1998.
[RFC2710] S. Deering, W. Fenner, B. Haberman, Multicast Listener Discovery (MLD) for IPv6, RFC 2710, October 1999.
[RFC3315] R. Droms, J. Bound, B. Volz, T. Lemon, C. Perkins, M. Carney, Dynamic Host Configuration Protocol for IPv6 (DHCPv6), RFC 3315, July 2003.

[RFC3590] b. Haberman, Source Address Selection for the Multicast Listener Discovery (MLD) Protocol, RFC 3590, September 2003.

[RFC 3810] R. Vida, L. Costa, Multicast Listener Discovery Version 2 (MLDv2) for IPv6, RFC 3810, June 2004.

Chapter 4

IPv6 Autoconfiguration Techniques

4.1 Introduction

Internet Protocol version 6 (IPv6) stateless address autoconfiguration (RFC 2462) specifies procedures by which a node may autoconfigure addresses based on router advertisements and the use of a valid lifetime to support renumbering of addresses on the Internet. In addition, the protocol interaction by which a node begins stateless or stateful autoconfiguration is specified. The Dynamic Host Control Protocol (DHCP) is one vehicle to perform stateful autoconfiguration; compatibility with stateless address autoconfiguration is a design requirement of DHCP [RFC3315].

4.2 Configuration Methods

As we have seen in previous chapters, the IPv6 protocol can use two address configuration methods: automatic configuration and manual configuration. Autoconfigured addresses exist in one or more of the states depicted in Figure 4.1: tentative, preferred, deprecated, valid (= preferred + deprecated), and invalid. IPv6 nodes (hosts and routers) automatically create unique link-local addresses for all local area network (LAN) interfaces that appear to be Ethernet interfaces. IPv6 hosts use received router advertisement messages to automatically configure the following parameters [MSD200401]:

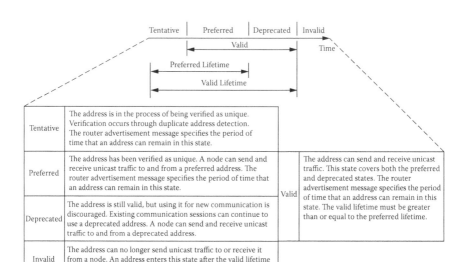

Tentative	The address is in the process of being verified as unique. Verification occurs through duplicate address detection. The router advertisement message specifies the period of time that an address can remain in this state.	
Preferred	The address has been verified as unique. A node can send and receive unicast traffic to and from a preferred address. The router advertisement message specifies the period of time that an address can remain in this state.	The address can send and receive unicast traffic. This state covers both the preferred and deprecated states. The router advertisement message specifies the period of time that an address can remain in this state. The valid lifetime must be greater than or equal to the preferred lifetime.
Deprecated	The address is still valid, but using it for new communication is discouraged. Existing communication sessions can continue to use a deprecated address. A node can send and receive unicast traffic to and from a deprecated address.	
Invalid	The address can no longer send unicast traffic to or receive it from a node. An address enters this state after the valid lifetime expires.	

Figure 4.1   Address states.

- A default router.
- The default setting for the hop limit field in the IPv6 header.
- The timers used in Neighbor Discovery processes.
- The maximum transfer unit (MTU) of the local link.
- The list of network prefixes that are defined for the link. Each network prefix contains both the IPv6 network prefix and its valid and preferred lifetimes. If indicated, a network prefix is combined with the interface identifier to create a stateless IPv6 address configuration for the receiving interface. A network prefix also defines the range of addresses for nodes on the local link.
- 6to4 addresses on a 6to4 tunneling interface for all public IPv4 addresses that are assigned to the computer (some implementations); see Chapter 6 for details of the 6to4 mechanism.
- Intrasite Automatic Tunnel Addressing Protocol (ISATAP) addresses on an automatic interface for all IPv4 addresses that are assigned to the computer (some implementations); see Chapter 6 for details of the ISATAP mechanism.
- The stack to query for IPv6 ISATAP routers in an IPv4 environment (some implementations).
- Routes to off-link prefixes if the off-link address prefix is advertised by a router (some implementations).

DHCP is not utilized in IPv6 to configure a link-local scope IP address. The link-local scope of an IPv6 addresses is always configured automatically. Addresses with other scopes, such as site local and global, are configured by router advertisements.

Specifically, unique link-local addresses are automatically configured for each interface on each IPv6 node (host or router). To communicate with IPv6 nodes that are not on attached links, the host must have additional site-local or global unicast addresses. Additional addresses for hosts are obtained from router advertisements; additional addresses for routers must be assigned manually. To communicate with IPv6 nodes on other network segments, IPv6 uses a default router. A default router is automatically assigned based on the receipt of a router advertisement. Alternately, one can add a default route to the IPv6 routing table. Note that one does not need to configure a default router for a network that consists of a single network segment.

The following sequence identifies the address autoconfiguration process for an IPv6 node, such as an IPv6-based Voice-over-IP (VoIP) phone.

1. A tentative link-local address is derived based on the link-local prefix of FE80::/64 and the 64-bit interface identifier.
2. Duplicate address detection is performed to verify the uniqueness of the tentative link-local address.
3. If duplicate address detection fails, one must manually configure the node.
   or
   If duplicate address detection succeeds, the tentative address is assumed to be valid and unique. The link-local address is initialized for the interface. The corresponding solicited-node multicast link-layer address is registered with the network adapter.

For an IPv6 host, address autoconfiguration continues as follows:

1. The host sends a router solicitation message.
2. If a router advertisement message is received, the configuration information that is included in the message is set on the host.
3. For each stateless autoconfiguration address prefix that is included, the following process occurs: The address prefix and the appropriate 64-bit interface identifier are used to derive a tentative address.
4. Duplicate address detection is used to verify the uniqueness of the tentative address. If the tentative address is in use, the address is not initialized for the interface. If the tentative address is not in use, the address is initialized. This includes setting the valid and preferred lifetimes based on information included in the router advertisement message.

Other configuration processes are shown in Table 4.1 [MSD200401].

4.3 Dynamic Host Configuration Protocol for IPv6

The Dynamic Host Control Protocol for IPv6 (DHCPv6 or more simply DHCP) enables DHCP servers to pass configuration parameters such as IPv6 network addresses to IPv6 nodes. DHCP provides both robust stateful autoconfiguration and autoregistration of DNS host names.

Table 4.1 Configurations of Interest

Configuration	Description
Single subnet with link-local addresses	This configuration supports the installation of the IPv6 protocol on at least two nodes on the same network segment without intermediate routers.
IPv6 traffic between nodes on different subnets of an IPv6 internetwork	This configuration includes two separate network segments (also known as links or subnets) and an IPv6-capable router that connects the network segments and forwards IPv6 protocol data units (PDUs) between the hosts.
IPv6 traffic between nodes on different subnets of an IPv4 internetwork	This configuration supports IPv6 traffic that is carried as the payload of an IPv4 PDU (treating the IPv4 infrastructure as an IPv6 link layer) without the deployment of IPv6 routers.
IPv6 traffic between nodes in different sites across the Internet	This configuration supports the 6to4 tunneling technique. (See Chapter 6 for details on the 6to4 mechanism.) The IPv6 traffic is encapsulated with an IPv4 header before it is sent over an IPv4 internetwork such as the Internet.

DHCP offers the capability of automatic allocation of reusable network addresses and additional configuration flexibility. This protocol is a stateful counterpart to IPv6 stateless address autoconfiguration [RFC2462] and can be used separately or concurrently with it to obtain configuration parameters. DHCP is a client/server protocol that provides managed configuration of devices. DHCP can provide a device with addresses assigned by a DHCP server and other configuration information. The operational models and relevant configuration information for DHCPv4 and DHCPv6 are significantly different [RFC3315].

Clients and servers exchange DHCP messages using UDP. The client uses a link-local address or an address determined through other mechanisms for transmitting and receiving DHCP messages. DHCP servers receive messages from clients using a reserved, link-scoped multicast address. A DHCP client transmits most messages to this reserved multicast address so that the client need not be configured with the address or addresses of DHCP servers [RFC3315].

DHCP makes use of the following multicast addresses:

- All_DHCP_Relay_Agents_and_Servers (FF02::1:2). A link-scoped multicast address used by a client to communicate with neighboring (i.e., on-link) relay agents and servers. All servers and relay agents are members of this multicast group.
- All_DHCP_Servers (FF05::1:3). A site-scoped multicast address used by a relay agent to communicate with servers either because the relay agent wants to send messages to all servers or because it does not know the unicast addresses of the servers. Note that for a relay agent to use this address, it must have an address of sufficient scope to be reachable by the servers. All servers within the site are members of this multicast group.

To allow a DHCP client to send a message to a DHCP server that is not attached to the same link, a DHCP relay agent on the client's link will relay messages between the client and server. The operation of the relay agent is transparent to the client, and the discussion of message exchanges in the remainder of this section omits the description of message relaying by relay agents. Once the client has determined the address of a server, it may under some circumstances send messages directly to the server using unicast [RFC3315].

The following list provides basic DHCP terminology:

- Appropriate to the link: An address is appropriate to the link when the address is consistent with the DHCP server's knowledge of the network topology, prefix assignment, and address assignment policies.
- Identity association (IA): A collection of addresses assigned to a client. Each IA has an associated identity association identifier (IAID). A client may have more than one IA assigned to it; for example, one for each of its interfaces. Each IA holds one type of address; for example, an identity association for temporary addresses (IA_TA) holds temporary addresses (see the identity association for temporary addresses entry). Throughout this section, IA is used to refer to an identity association without identifying the type of addresses in the IA.
- DUID: A DHCP unique identifier for a DHCP participant; each DHCP client and server has exactly one DUID.
- Binding: A binding (or client binding) is a group of server data records containing the information the server has about the addresses in an IA or configuration information explicitly assigned to the client. Configuration information that has been returned to a client through a policy (e.g., the information returned to all clients on the same link) does not require a binding. A binding containing information about an IA is indexed by the tuple <DUID, IA-type, IAID> (where IA-type is the type of address in the IA; for

example, temporary). A binding containing configuration information for a client is indexed by <DUID>.

■ Configuration parameter: An element of the configuration information set on the server and delivered to the client using DHCP. Such parameters may be used to carry information to be used by a node to configure its network subsystem and enable communication, for example, on a link or internetwork.

■ DHCP client (or client): A node that initiates requests on a link to obtain configuration parameters from one or more DHCP servers.

■ DHCP domain: A set of links managed by DHCP and operated by a single administrative entity.

■ DHCP realm: A name used to identify the DHCP administrative domain from which a DHCP authentication key was selected.

■ DHCP relay agent (or relay agent): A node that acts as an intermediary to deliver DHCP messages between clients and servers and is on the same link as the client.

■ DHCP server (or server): A node that responds to requests from clients and may or may not be on the same link as the client(s).

■ Identity association identifier (IAID): An identifier for an IA, chosen by the client. Each IA has an IAID, which is chosen to be unique among all IAIDs for IAs belonging to that client.

■ Identity association for nontemporary addresses (IA_NA): An IA that carries assigned addresses that are not temporary addresses (see the identity association for temporary addresses entry).

■ Identity association for temporary addresses (IA_TA): An IA that carries temporary addresses.

■ Message: A unit of data carried as the payload of a UDP datagram, exchanged among DHCP servers, relay agents and clients.

■ Reconfigure key: A key supplied to a client by a server used to provide security for reconfigure messages.

■ Relaying: A DHCP relay agent relays DHCP messages between DHCP participants.

■ Transaction ID: An opaque value used to match responses with replies initiated either by a client or server.

Clients listen for DHCP messages on UDP port 546. Servers and relay agents listen for DHCP messages on UDP port 547. DHCP defines and makes use of the following message types [RFC3315]:

1. Solicit: A client sends a solicit message to locate servers.
2. Advertise: A server sends an advertise message to indicate that it is available for DHCP service; this is in response to a solicit message received from a client.
3. Request: A client sends a request message to request configuration parameters, including IP addresses, from a specific server.

4. Confirm: A client sends a confirm message to any available server to determine whether the addresses it was assigned are still appropriate to the link to which the client is connected.

5. Renew: A client sends a renew message to the server that originally provided the client's addresses and configuration parameters to extend the lifetimes on the addresses assigned to the client and to update other configuration parameters.

6. Rebind: A client sends a rebind message to any available server to extend the lifetimes on the addresses assigned to the client and to update other configuration parameters; this message is sent after a client receives no response to a renew message.

7. Reply: A server sends a reply message containing assigned addresses and configuration parameters in response to a solicit, request, renew, rebind message received from a client. A server sends a reply message containing configuration parameters in response to an information-request message. A server sends a reply message in response to a confirm message confirming or denying that the addresses assigned to the client are appropriate to the link to which the client is connected. A server sends a reply message to acknowledge receipt of a release or decline message.

8. Release: A client sends a release message to the server that assigned addresses to the client to indicate that the client will no longer use one or more of the assigned addresses.

9. Decline: A client sends a decline message to a server to indicate that the client has determined that one or more addresses assigned by the server are already in use on the link to which the client is connected.

10. Reconfigure: A server sends a reconfigure message to a client to inform the client that the server has new or updated configuration parameters, and that the client is to initiate a renew/reply or information-request/reply transaction with the server to receive the updated information.

11. Information-request: A client sends an information-request message to a server to request configuration parameters without the assignment of any IP addresses to the client.

12. Relay-forward: A relay agent sends a relay-forward message to relay messages to servers, either directly or through another relay agent. The received message, either a client message or a relay-forward message from another relay agent, is encapsulated in an option in the relay-forward message.

13. Relay-reply: A server sends a relay-reply message to a relay agent containing a message that the relay agent delivers to a client. The relay-reply message may be relayed by other relay agents for delivery to the destination relay agent. The server encapsulates the client message as an option in the relay-reply message, which the relay agent extracts and relays to the client.

References

[MSD200401] Microsoft Corporation, MSDN Library, Internet Protocol, 2004, http://msdn.microsoft.com.

[RFC2462] S. Thomson, T. Narten, IPv6 Stateless Address Autoconfiguration, RFC 2462, December 1998.

[RFC3315] R. Droms, Ed., J. Bound, B. Volz, T. Lemon, C. Perkins, and M. Carney, Dynamic Host Configuration Protocol for IPv6 (DHCPv6), RFC 3315, July 2003.

Chapter 5

IPv6 and Related Protocols (Details)

We introduced a number of basic Internet Protocol version 6 (IPv6) concepts in previous chapters. This chapter focuses on a more formal description of IPv6. The discussion is based on Internet Engineering Task Force (IETF) [RFC2460]. There is an extensive body of technical research literature on this topic (as documented in Appendix B).

5.1 Introduction

IPv6 is a new version of the Internet Protocol, designed as the successor to IP version 4 (IPv4) described in [RFC791]. The changes from IPv4 to IPv6 fall primarily in the following categories:

- Expanded addressing capabilities. IPv6 increases the IP address size from 32 bits to 128 bits to support more levels of addressing hierarchy, a much greater number of addressable nodes, and a simpler autoconfiguration addressing scheme. The scalability of multicast routing is improved by adding a scope field to multicast addresses, and a new type of address called an *anycast address* is defined, to be used to send a packet to any one of a group of nodes.
- Header format simplification. Some IPv4 header fields have been dropped or made optional to reduce the common-case processing cost of packet handling and to limit the bandwidth cost of the IPv6 header.

- Improved support for extensions and options. Changes in the way IP header options are encoded allows for more efficient forwarding, less-stringent limits on the length of options, and greater flexibility for introducing new options in the future.
- Flow labeling capability. A new capability is added to enable the labeling of packets belonging to particular traffic "flows" for which the sender requests special handling, such as nondefault quality of service or "real-time" service.
- Authentication and privacy capabilities. Extensions to support authentication, data integrity, and (optional) data confidentiality are specified for IPv6.

RFC 2460 specifies the basic IPv6 header and the initially defined IPv6 extension headers and options. It also discusses packet size issues, the semantics of flow labels and traffic classes, and the effects of IPv6 on upper-layer protocols. The format and semantics of IPv6 addresses are specified separately in RFC 2373 (now obsoleted by [RFC4291]). The IPv6 version of the Internet Control Messaging Protocol (ICMP), which all IPv6 implementations are required to include, is specified in ICMPv6 [RFC2483]. Developers should refer directly to all relevant IETF RFCs for normative guidelines.

5.2 Terminology

The following nomenclature is used in the standard.

- Node: A device that implements IPv6.
- Router: A node that forwards IPv6 packets not explicitly addressed to itself (see note below).
- Host: Any node that is not a router (see note below).
- Upper layer: A protocol layer immediately above IPv6. Examples are transport protocols such as Transmission Control Protocol (TCP) and User Datagram Protocol (UDP), control protocols such as ICMP, routing protocols such as Open Shortest Path First (OSPF), and Internet or lower-layer protocols being "tunneled" over (i.e., encapsulated in) IPv6 such as Internet packet exchange (IPX), AppleTalk, or IPv6 itself.
- Link: A communication facility or medium over which nodes can communicate at the link layer, that is, the layer immediately below IPv6. Examples are Ethernets (simple or bridged), Point-to-Point Protocol (PPP) links, X.25, Frame Relay, or Asynchronous Transfer Mode (ATM) networks, and Internet (or higher) layer tunnels, such as tunnels over IPv4 or IPv6 itself.
- Neighbors: Nodes attached to the same link.
- Interface: A node's attachment to a link.
- Address: An IPv6 layer identifier for an interface or a set of interfaces.

- Packet: An IPv6 header plus payload.
- Link maximum transfer unit (MTU): The maximum transfer unit (i.e., maximum packet size in octets) that can be conveyed over a link.
- Path MTU: The minimum link MTU of all the links in a path between a source node and a destination node.

Note that it is possible, although unusual, for a device with multiple interfaces to be configured to forward non-self-destined packets arriving from some set (fewer than all) of its interfaces and to discard non-self-destined packets arriving from its other interfaces. Such a device must obey the protocol requirements for routers when receiving packets from, and interacting with neighbors, the former (forwarding) interfaces. It must obey the protocol requirements for hosts when receiving packets from, and interacting with neighbors over, the latter (nonforwarding) interfaces.

5.3 IPv6 Header Format

```
+-+-+-+-+-+-+-+-+-+-+-+-+-+-+-+-+-+-+-+-+-+-+-+-+-+-+-+-+-+-+-+-+
|Version| Traffic Class |              Flow Label               |
+-+-+-+-+-+-+-+-+-+-+-+-+-+-+-+-+-+-+-+-+-+-+-+-+-+-+-+-+-+-+-+-+
|         Payload Length        | Next Header   |   Hop Limit   |
+-+-+-+-+-+-+-+-+-+-+-+-+-+-+-+-+-+-+-+-+-+-+-+-+-+-+-+-+-+-+-+-+
|                                                               |
+                                                               +
|                                                               |
+                     Source Address                            +
|                                                               |
+                                                               +
|                                                               |
+-+-+-+-+-+-+-+-+-+-+-+-+-+-+-+-+-+-+-+-+-+-+-+-+-+-+-+-+-+-+-+-+
|                                                               |
+                                                               +
|                                                               |
+                   Destination Address                         +
|                                                               |
+                                                               +
|                                                               |
+-+-+-+-+-+-+-+-+-+-+-+-+-+-+-+-+-+-+-+-+-+-+-+-+-+-+-+-+-+-+-+-+
```

Figure 5.1   The IPv6 header format.

Figure 5.1 depicts the IPv6 Header format. The fields in the header have the following meanings:

Version: Four-bit Internet Protocol version number = 6.

Traffic Class: Eight-bit traffic class field (See Section 5.8).

Flow Label: A 20-bit flow label (see Section 5.7).

Payload Length: A 16-bit unsigned integer; length of the IPv6 payload (i.e., the rest of the packet following this IPv6 header) in octets. (Note that any extension headers [Section 5.5] present are considered part of the payload, i.e., are included in the length count.)

Next Header: Eight-bit selector. Identifies the type of header immediately following the IPv6 header. Uses the same values as the IPv4 protocol field.

Hop Limit: Eight-bit unsigned integer. Decremented by one by each node that forwards the packet. The packet is discarded if hop limit is decremented to zero.

Source Address: The 128-bit address of the originator of the packet. This is covered later in more detail.

Destination Address: The 128-bit address of the intended recipient of the packet (possibly not the ultimate recipient if a routing header is present).

5.4 IPv6 Extension Headers

In IPv6, optional Internet layer information is encoded in separate headers that may be placed between the IPv6 header and the upper-layer header in a packet. There are a small number of such extension headers, each identified by a distinct next header value. As illustrated in the examples of Figure 5.2, an IPv6 packet may carry zero, one, or more extension headers, each identified by the next header field of the preceding header.

With one exception, extension headers are not examined or processed by any node along a packet's delivery path until the packet reaches the node (or each of the set of nodes, in the case of multicast) identified in the destination address field of the IPv6 header. There, normal demultiplexing on the next header field of the IPv6 header invokes the module to process the first extension header or the upper-layer header if no extension header is present. The contents and semantics of each extension header determine whether to proceed to the next header. Therefore, extension headers must be processed strictly in the order they appear in the packet. A receiver must not, for example, scan through a packet looking for a particular kind of extension header and process that header prior to processing all preceding ones.

The exception referred to in the preceding paragraph is the Hop-by-Hop Options header, which carries information that must be examined and processed by every node along a packet's delivery path, including the source and destination nodes. The Hop-by-Hop Options header, when present, must immediately follow the IPv6 header. Its presence is indicated by the value zero in the next header field of the IPv6 header.

If, as a result of processing a header, a node is required to proceed to the next header but the next header value in the current header is unrecognized by the node,

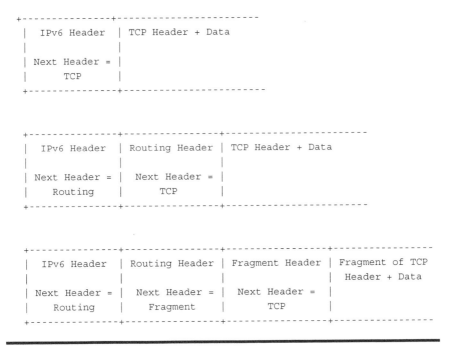

Figure 5.2 Examples of extension headers.

it should discard the packet and send an ICMP parameter problem message to the source of the packet, with an ICMP code value of one (unrecognized next header type encountered) and the ICMP pointer field containing the offset of the unrecognized value within the original packet. The same action should be taken if a node encounters a next header value of zero in any header other than an IPv6 header.

Each extension header is an integer multiple of eight octets long to retain eight-octet alignment for subsequent headers. Multi-octet fields within each extension header are aligned on their natural boundaries; that is, fields of width n octets are placed at an integer multiple of n octets from the start of the header, for n = 1, 2, 4, or 8.

A full implementation of IPv6 includes implementation of the following extension headers:

Hop-by-Hop Options
Routing (Type 0)
Fragment
Destination Options
Authentication
Encapsulating Security Payload

The first four are specified in [RFC2460]; the last two are specified in [RFC2402] and [RFC2406], respectively.

5.4.1 Extension Header Order

When more than one extension header is used in the same packet, it is recommended that those headers appear in the following order:

IPv6 header
Hop-by-Hop Options header
Destination Options header (Note 1)
Routing header
Fragment header
Authentication header (Note 2)
Encapsulating Security Payload header (Note 2)
Destination Options header (Note 3)
Upper-layer header

Note 1: For options to be processed by the first destination that appears in the IPv6 destination address field plus subsequent destinations listed in the routing header.

Note 2: Additional recommendations regarding the relative order of the Authentication and Encapsulating Security Payload headers are given in [RFC2406].

Note 3: For options to be processed only by the final destination of the packet.

Each extension header should occur at most once, except for the Destination Options header, which should occur at most twice (once before a Routing header and once before the Upper-layer header).

If the Upper-layer header is another IPv6 header (in the case of IPv6 being tunneled over or encapsulated in IPv6), it may be followed by its own extension headers, which are separately subject to the same ordering recommendations.

If and when other extension headers are defined, their ordering constraints relative to the above-listed headers must be specified.

IPv6 nodes must accept and attempt to process extension headers in any order and occurring any number of times in the same packet, except for the Hop-by-Hop Options header, which is restricted to appear immediately after an IPv6 header only. Nonetheless, it is strongly advised that sources of IPv6 packets adhere to the above-recommended order until and unless subsequent specifications revise that recommendation.

5.4.2 Options

Two of the currently defined extension headers — the Hop-by-Hop Options header and the Destination Options header — carry a variable number of type-length-value (TLV) encoded "options" of the format shown in Figure 5.3:

```
+-+-+-+-+-+-+-+-+-+-+-+-+-+-+-+-+- - - - - - - -
|  Option Type  |  Opt Data Len  |  Option Data
+-+-+-+-+-+-+-+-+-+-+-+-+-+-+-+-+- - - - - - - -
```

Figure 5.3   Extension header options.

Option Type: Eight-bit identifier of the type of option

Opt Data Len: Eight-bit unsigned integer; length of the option data field of this option, in octets

Option Data: Variable-length field; option-type-specific data

The sequence of options within a header must be processed strictly in the order they appear in the header; a receiver must not, for example, scan through the header looking for a particular kind of option and process that option prior to processing all preceding ones.

The Option Type identifiers are internally encoded such that their highest-order two bits specify the action that must be taken if the processing IPv6 node does not recognize the Option Type:

- 00: Skip over this option and continue processing the header.
- 01: Discard the packet.
- 10: Discard the packet and, regardless of whether the packet's destination address was a multicast address, send an ICMP parameter problem, Code 2, message to the packet's source address, pointing to the unrecognized Option Type.
- 11: Discard the packet and, only if the packet's destination address was not a multicast address, send an ICMP parameter problem, Code 2, message to the packet's source address, pointing to the unrecognized Option Type.

The third-highest-order bit of the Option Type specifies whether the option data of that option can change en route to the packet's final destination. When an Authentication header is present in the packet, for any option with data that may change en route, its entire option data field must be treated as zero-valued octets when computing or verifying the packet's authenticating value:

- 0: Option data does not change en route.
- 1: Option data may change en route.

The three high-order bits described above are to be treated as part of the Option Type, not independent of the Option Type. That is, a particular option is identified by a full eight-bit Option Type, not just the low-order five bits of an Option Type.

The same Option Type numbering space is used for both the Hop-by-Hop Options header and the Destination Options header. However, the specification of a particular option may restrict its use to only one of those two headers.

Individual options may have specific alignment requirements to ensure that multi-octet values within option data fields fall on natural boundaries. The alignment requirement of an option is specified using the notation xn+y, meaning the option type must appear at an integer multiple of x octets from the start of the header plus y octets. For example:

2n means any two-octet offset from the start of the header.
8n+2 means any eight-octet offset from the start of the header plus two octets.

There are two padding options that are used when necessary to align subsequent options and to pad out the containing header to a multiple of eight octets in length. These padding options must be recognized by all IPv6 implementations:

```
Pad1 option (alignment requirement: none)

      +-+-+-+-+-+-+-+-+
      |       0       |
      +-+-+-+-+-+-+-+-+
```

The Pad1 option is used to insert one octet of padding into the options area of a header. Note that the format of the Pad1 option is a special case — it does not have length and value fields.

If more than one octet of padding is required, the PadN option, described next, should be used, rather than multiple Pad1 options.

```
PadN option (alignment requirement: none)

      +-+-+-+-+-+-+-+-+-+-+-+-+-+-+-+-+-  - - - - - - -
      |       1       | Opt Data Len |  Option Data
      +-+-+-+-+-+-+-+-+-+-+-+-+-+-+-+-+-  - - - - - - -
```

The PadN option is used to insert two or more octets of padding into the options area of a header. For N octets of padding, the Opt Data Len field contains the value N-2, and the option data consists of N-2 zero-valued octets.

Section 5.10 contains formatting guidelines for designing new options.

5.4.3 Hop-by-Hop Options Header

The Hop-by-Hop Options header is used to carry optional information that must be examined by every node along a packet's delivery path. The Hop-by-Hop Options header is identified by a next header value of zero in the IPv6 header and has the format of Figure 5.4.

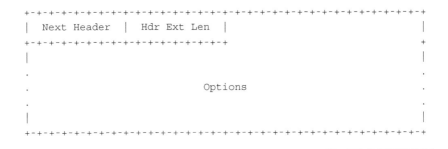

Figure 5.4   Hop-by-Hop Options header.

The fields are as follows:

Next Header: Eight-bit selector. Identifies the type of header immediately following the Hop-by-Hop Options header. Uses the same values as the IPv4 protocol field [RFC3232].

Hdr Ext Len: Eight-bit unsigned integer. Length of the Hop-by-Hop Options header in eight-octet units, not including the first eight octets.

Options: Variable-length field, of length such that the complete Hop-by-Hop Options header is an integer multiple of eight octets long. Contains one or more TLV-encoded options, as described in Section 5.5.2.

The only Hop-by-Hop Options defined in this RFC are the Pad1 and PadN options specified in Section 5.5.2.

5.4.4 Routing Header

The Routing header is used by an IPv6 source to list one or more intermediate nodes to be "visited" on the way to a packet's destination. This function is very similar to IPv4's loose source and record route option. The Routing header is identified by a next header value of 43 in the immediately preceding header and has the format of Figure 5.5.

The fields are as follows:

Next Header: Eight-bit selector. Identifies the type of header immediately following the routing header. Uses the same values as the IPv4 protocol field [RFC3232].

Hdr Ext Len: Eight-bit unsigned integer. Length of the routing header in eight-octet units, not including the first eight octets.

Routing Type: Eight-bit identifier of a particular routing header variant.

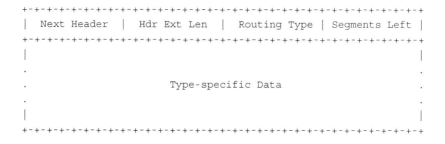

Figure 5.5 Routing header.

Segments Left: Eight-bit unsigned integer. Number of route segments remaining (i.e., number of explicitly listed intermediate nodes still to be visited before reaching the final destination).

Type-specific data: Variable-length field, of format determined by the routing type and of length such that the complete routing header is an integer multiple of eight octets long.

If, while processing a received packet, a node encounters a Routing header with an unrecognized Routing Type value, the required behavior of the node depends on the value of the Segments Left field, as follows:

If Segments Left is zero, the node must ignore the Routing header and proceed to process the next header in the packet, which has a type that is identified by the next header field in the routing header.

If Segments Left is nonzero, the node must discard the packet and send an ICMP parameter problem, Code 0, message to the packet's source address, pointing to the unrecognized Routing Type.

If, after processing a Routing header of a received packet, an intermediate node determines that the packet is to be forwarded on to a link with a link MTU that is less than the size of the packet, the node must discard the packet and send an ICMP packet too big message to the packet's source address.

The Type 0 Routing header has the format shown in Figure 5.6. The fields are as follows:

Next Header: Eight-bit selector. Identifies the type of header immediately following the routing header. Uses the same values as the IPv4 protocol field [RFC3232].

Hdr Ext Len: Eight-bit unsigned integer. Length of the Routing header in eight-octet units, not including the first eight octets. For the Type 0 Routing header, Hdr Ext Len is equal to two times the number of addresses in the header.

Routing Type: Zero.

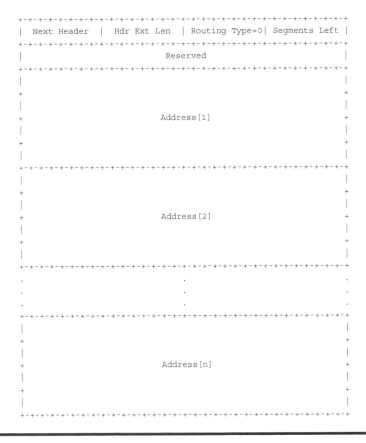

Figure 5.6   Type 0 routing header.

Segments Left: Eight-bit unsigned integer. Number of route segments remaining (i.e., number of explicitly listed intermediate nodes still to be visited before reaching the final destination).

Reserved: A 32-bit reserved field. This is initialized to zero for transmission and ignored on reception.

Address[1, 2, … , n]: Vector of 128-bit addresses, numbered 1 to n.

Multicast addresses must not appear in a routing header of Type 0 or in the IPv6 Destination Address field of a packet carrying a Routing header of Type 0.

A Routing header is not examined or processed until it reaches the node identified in the Destination Address field of the IPv6 header. In that node, dispatching on the next header field of the immediately preceding header causes the Routing header module to be invoked, which in the case of Routing Type 0, performs the following algorithm:

```
if Segments Left = 0 {
   proceed to process the next header in the packet, whose
    type is identified by the Next Header field in the
    Routing header
}
else if Hdr Ext Len is odd {
      send an ICMP Parameter Problem, Code 0, message to
       the Source Address, pointing to the Hdr Ext Len
       field, and discard the packet
}
else {
   compute n, the number of addresses in the Routing
    header, by dividing Hdr Ext Len by 2

   if Segments Left is greater than n {
      send an ICMP Parameter Problem, Code 0, message to
       the Source Address, pointing to the Segments Left
       field, and discard the packet
   }
   else {
      decrement Segments Left by 1;
      compute i, the index of the next address to be
       visited in the address vector, by subtracting
       Segments Left from n

      if Address [i] or the IPv6 Destination Address is
       multicast {
         discard the packet
      }
      else {
         swap the IPv6 Destination Address and Address[i]

         if the IPv6 Hop Limit is less than or equal to 1 {
            send an ICMP Time Exceeded -- Hop Limit
             Exceeded in Transit message to the Source
             Address and discard the packet
         }
         else {
            decrement the Hop Limit by 1

            resubmit the packet to the IPv6 module for
             transmission to the new destination
         }
      }
   }
}
```

As an example of the effects of the above algorithm, consider the case of a source node S sending a packet to destination node D using a Routing header to cause the packet to be routed via intermediate nodes I1, I2, and I3. The values of the relevant IPv6 header and routing header fields on each segment of the delivery path would be as follows:

As the packet travels from S to I1:

```
Source Address = S                Hdr Ext Len = 6
Destination Address = I1          Segments Left = 3
                                  Address[1] = I2
                                  Address[2] = I3
                                  Address[3] = D
```

As the packet travels from I1 to I2:

```
Source Address = S                Hdr Ext Len = 6
Destination Address = I2          Segments Left = 2
                                  Address[1] = I1
                                  Address[2] = I3
                                  Address[3] = D
```

As the packet travels from I2 to I3:

```
Source Address = S                Hdr Ext Len = 6
Destination Address = I3          Segments Left = 1
                                  Address[1] = I1
                                  Address[2] = I2
                                  Address[3] = D
```

As the packet travels from I3 to D:

```
Source Address = S                Hdr Ext Len = 6
Destination Address = D           Segments Left = 0
                                  Address[1] = I1
                                  Address[2] = I2
                                  Address[3] = I3
```

5.4.5 Fragment Header

The Fragment header is used by an IPv6 source to send a packet larger than would fit in the path MTU to its destination. (Note that, unlike IPv4, fragmentation in IPv6 is performed only by source nodes, not by routers along a packet's delivery path; see Section 5.6.) The Fragment header is identified by a next header value of

```
+-+-+-+-+-+-+-+-+-+-+-+-+-+-+-+-+-+-+-+-+-+-+-+-+-+-+-+-+-+-+-+-+
|   Next Header   |    Reserved    |       Fragment Offset       |Res|M|
+-+-+-+-+-+-+-+-+-+-+-+-+-+-+-+-+-+-+-+-+-+-+-+-+-+-+-+-+-+-+-+-+
|                          Identification                        |
+-+-+-+-+-+-+-+-+-+-+-+-+-+-+-+-+-+-+-+-+-+-+-+-+-+-+-+-+-+-+-+-+
```

Figure 5.7   Fragment header.

44 in the immediately preceding header and has the format shown in Figure 5.7. The fields are as follows:

Next Header: Eight-bit selector. Identifies the initial header type of the fragmentable part of the original packet (defined below). Uses the same values as the IPv4 protocol field (RFC 1700).

Reserved: Eight-bit reserved field. This field is initialized to zero for transmission and ignored on reception.

Fragment Offset: A 13-bit unsigned integer. The offset, in eight-octet units, of the data following this header, relative to the start of the fragmentable part of the original packet.

Res: Two-bit reserved field. This field is initialized to zero for transmission and ignored on reception.

M flag: 1 = more fragments; 0 = last fragment.

Identification: This is 32 bits. See description below.

To send a packet that is too large to fit in the MTU of the path to its destination, a source node may divide the packet into fragments and send each fragment as a separate packet to be reassembled at the receiver.

For every packet that is to be fragmented, the source node generates an Identification value. The Identification must be different from that of any other fragmented packet sent recently[1] with the same source address and destination address. If a Routing header is present, the Destination Address of concern is that of the final destination.

The initial, large, unfragmented packet is referred to as the "original packet," and it is considered to consist of two parts, as seen in Figure 5.8.

---

[1] *Recently* means within the maximum likely lifetime of a packet, including transit time from source to destination and time spent awaiting reassembly with other fragments of the same packet. However, it is not required that a source node know the maximum packet lifetime. Rather, it is assumed that the requirement can be met by maintaining the identification value as a simple, 32-bit, "wrap-around" counter, incremented each time a packet must be fragmented. It is an implementation choice whether to maintain a single counter for the node or multiple counters, such as one for each of the node's possible source addresses or one for each active (source address, destination address) combination.

```
= = = == == = = == == = = == == = =

Original Packet:

    +-----------------+--------------------//----------------------+
    |  Unfragmentable |            Fragmentable                    |
    |      Part       |                Part                        |
    +-----------------+--------------------//----------------------+


= = = == == = = == == = = == == = =
```

Figure 5.8   Original packet.

The unfragmentable part consists of the IPv6 header plus any extension headers
that must be processed by nodes en route to the destination, that is, all headers
up to and including the Routing header if present, or else the Hop-by-Hop
Options header if present, or else no extension headers.

The fragmentable part consists of the rest of the packet, that is, any extension
headers that need to be processed only by the final destination nodes, plus
the Upper-layer header and data.

The fragmentable part of the original packet is divided into fragments, with
each, except possibly the last ("rightmost") one, an integer multiple of eight octets
long. The fragments are transmitted in separate fragment packets as illustrated in
Figure 5.9. Each fragment packet comprises three components.

The unfragmentable part of the original packet, with the payload length of the
original IPv6 header changed to contain the length of this fragment packet
only (excluding the length of the IPv6 header itself) and the Next Header field
of the last header of the unfragmentable part changed to 44.

A Fragment header containing the following:

The next header value that identifies the first header of the fragmentable part
of the original packet.

A Fragment Offset containing the offset of the fragment, in eight-octet units,
relative to the start of the fragmentable part of the original packet. The
Fragment Offset of the first ("leftmost") fragment is zero.

An M flag value of zero if the fragment is the last (rightmost) one, else an M
flag value of one.

The Identification value generated for the original packet.

The fragment itself.

The lengths of the fragments must be chosen such that the resulting fragment
packets fit within the MTU of the path to the packets' destinations.

```
Original Packet:

+------------------+--------------+--------------+--//--+----------+
|  Unfragmentable  |    First     |    Second    |      |   Last   |
|       Part       |   Fragment   |   Fragment   | .... | Fragment |
+------------------+--------------+--------------+--//--+----------+

Fragment Packets:

+------------------+--------+--------------+
|  Unfragmentable  |Fragment|    First     |
|       Part       | Header |   Fragment   |
+------------------+--------+--------------+


+------------------+--------+--------------+
|  Unfragmentable  |Fragment|    Second    |
|       Part       | Header |   Fragment   |
+------------------+--------+--------------+
                        o
                        o
                        o
+------------------+--------+----------+
|  Unfragmentable  |Fragment|   Last   |
|       Part       | Header | Fragment |
+------------------+--------+----------+
```

Figure 5.9   Fragmentable parts.

```
= = = == == = = == == = = == == = =
Reassembled Original Packet:

+------------------+----------------------//----------------------+
|  Unfragmentable  |                 Fragmentable                 |
|       Part       |                     Part                     |
+------------------+----------------------//----------------------+

= = = == == = = == == = = == == = =
```

Figure 5.10   Reassembled original packet.

At the destination, fragment packets are reassembled into their original, unfragmented form, as illustrated in Figure 5.10. The following rules govern reassembly:

■ An original packet is reassembled only from fragment packets that have the same source address, destination address, and fragment identification.
■ The unfragmentable part of the reassembled packet consists of all headers up to, but not including, the Fragment header of the first fragment packet (that

is, the packet with a Fragment Offset that is zero), with the following two changes:

The Next Header field of the last header of the unfragmentable part is obtained from the Next Header field of the first fragment's fragment header.

The Payload Length of the reassembled packet is computed from the length of the unfragmentable part and the length and offset of the last fragment. For example, a formula for computing the Payload Length of the reassembled original packet is:

$$PL.orig = PL.first - FL.first - 8 + (8 * FO.last) + FL.last$$

where

PL.orig = Payload Length field of reassembled packet.

PL.first = Payload Length field of first fragment packet.

FL.first = length of fragment following Fragment header of first fragment packet.

FO.last = Fragment Offset field of Fragment header of last fragment packet.

FL.last = length of fragment following Fragment header of last fragment packet.

■ The fragmentable part of the reassembled packet is constructed from the fragments following the Fragment headers in each of the fragment packets. The length of each fragment is computed by subtracting from the packet's payload length the length of the headers between the IPv6 header and fragment itself; its relative position in the fragmentable part is computed from its Fragment Offset value.

■ The Fragment header is not present in the final, reassembled packet.

The following error conditions may arise when reassembling fragmented packets:

■ If insufficient fragments are received to complete reassembly of a packet within 60 seconds of the reception of the first-arriving fragment of that packet, reassembly of that packet must be abandoned, and all the fragments that have been received for that packet must be discarded. If the first fragment (i.e., the one with a Fragment Offset of zero) has been received, an ICMP time exceeded — fragment reassembly time exceeded message should be sent to the source of that fragment.

■ If the length of a fragment, as derived from the fragment packet's Payload Length field, is not a multiple of eight octets and the M flag of that fragment is one, then that fragment must be discarded, and an ICMP parameter problem, Code 0, message should be sent to the source of the fragment, pointing to the Payload Length field of the fragment packet.

- If the length and offset of a fragment are such that the payload length of the packet reassembled from that fragment would exceed 65,535 octets, then that fragment must be discarded and an ICMP parameter problem, Code 0, message should be sent to the source of the fragment, pointing to the Fragment Offset field of the fragment packet.

The following conditions are not expected to occur but are not considered errors if they do:

- The number and content of the headers preceding the Fragment header of different fragments of the same original packet may differ. Whatever headers are present, preceding the Fragment header in each fragment packet, are processed when the packets arrive, prior to queueing the fragments for reassembly. Only those headers in the offset zero fragment packet are retained in the reassembled packet.
- The Next Header values in the Fragment headers of different fragments of the same original packet may differ. Only the value from the offset zero fragment packet is used for reassembly.

5.4.6 Destination Options Header

The Destination Options header is used to carry optional information that need be examined only by a packet's destination nodes. The Destination Options header is identified by a next header value of 60 in the immediately preceding header and has the format shown in Figure 5.11:

Next Header: Eight-bit selector. Identifies the type of header immediately following the destination options header. Uses the same values as the IPv4 protocol field [RFC3232].

Hdr Ext Len: Eight-bit unsigned integer. Length of the Destination Options header in eight-octet units, not including the first eight octets.

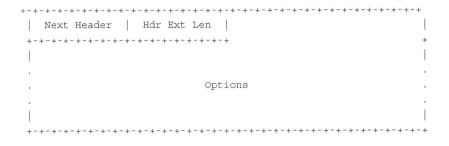

Figure 5.11   Destination options header.

Options: Variable-length field, of length such that the complete Destination Options header is an integer multiple of eight octets long. Contains one or more TLV-encoded options, as described in Section 5.5.2.

The only destination options defined in this RFC are the Pad1 and PadN options specified in Section 5.5.2.

Note that there are two possible ways to encode optional destination information in an IPv6 packet: either as an option in the Destination Options header or as a separate extension header. The Fragment header and the Authentication header are examples of the latter approach. Which approach can be used depends on what action is desired of a destination node that does not understand the optional information:

- If the desired action is for the destination node to discard the packet and, only if the packet's destination address is not a multicast address, send an ICMP unrecognized type message to the packet's source address, then the information may be encoded either as a separate header or as an option in the Destination Options header with an option Type that has the value 11 in its highest-order two bits. The choice may depend on such factors as which takes fewer octets or which yields better alignment or more efficient parsing.
- If any other action is desired, the information must be encoded as an option in the Destination Options header with an option Type that has the value 00, 01, or 10 in its highest-order two bits, specifying the desired action (see Section 5.5.2).

5.4.7 No Next Header

The value 59 in the next header field of an IPv6 header or any extension header indicates that there is nothing following that header. If the Payload Length field of the IPv6 header indicates the presence of octets past the end of a header with a next header field that contains 59, those octets must be ignored and passed on unchanged if the packet is forwarded.

5.5 Packet Size Issues

IPv6 requires that every link in the Internet have an MTU of 1280 octets or greater. On any link that cannot convey a 1280-octet packet in one piece, link-specific fragmentation and reassembly must be provided at a layer below IPv6.

Links that have a configurable MTU (e.g., PPP links defined in [RFC1661]) must be configured to have an MTU of at least 1280 octets. It is recommended that they be configured with an MTU of 1500 octets or greater to accommodate possible encapsulations (i.e., tunneling) without incurring IPv6 layer fragmentation.

From each link to which a node is directly attached, the node must be able to accept packets as large as that link's MTU.

It is strongly recommended that IPv6 nodes implement path MTU discovery [RFC1981] to discover and take advantage of path MTUs greater than 1280 octets. However, a minimal IPv6 implementation (e.g., in a boot read-only memory (ROM)) may simply restrict itself to sending packets no larger than 1280 octets and omit implementation of path MTU discovery.

To send a packet larger than a path's MTU, a node may use the IPv6 Fragment header to fragment the packet at the source and have it reassembled at the destinations. However, the use of such fragmentation is discouraged in any application that is able to adjust its packets to fit the measured path MTU (i.e., down to 1280 octets).

A node must be able to accept a fragmented packet that, after reassembly, is as large as 1500 octets. A node is permitted to accept fragmented packets that reassemble to more than 1500 octets. An upper-layer protocol or application that depends on IPv6 fragmentation to send packets larger than the MTU of a path should not send packets larger than 1500 octets unless it has assurance that the destination is capable of reassembling packets of that larger size.

In response to an IPv6 packet that is sent to an IPv4 destination (i.e., a packet that undergoes translation from IPv6 to IPv4), the originating IPv6 node may receive an ICMP packet too big message reporting a next-hop MTU less than 1280. In that case, the IPv6 node is not required to reduce the size of subsequent packets to less than 1280 but must include a fragment header in those packets so that the IPv6-to-IPv4 translating router can obtain a suitable identification value to use in resulting IPv4 fragments. Note that this means the payload may have to be reduced to 1232 octets (1280 minus 40 for the IPv6 header and 8 for the fragment header) and smaller still if additional extension headers are used.

5.6 Flow Labels

The 20-bit Flow Label field in the IPv6 header may be used by a source to label sequences of packets for which it requests special handling by the IPv6 routers, such as nondefault quality of service or real-time service. This aspect of IPv6 is still experimental to a large degree and subject to change as the requirements for flow support in the Internet become clearer ([RFC3697] and [RFC3595] provide some current thinking on the topic). Hosts or routers that do not support the functions of the Flow Label field are required to set the field to zero when originating a packet, pass the field on unchanged when forwarding a packet, and ignore the field when receiving a packet.

Section 5.9 describes the current intended semantics and usage of the Flow Label field.

5.7 Traffic Classes

The eight-bit Traffic Class field in the IPv6 header is available for use by originating nodes or forwarding routers to identify and distinguish between different classes or priorities of IPv6 packets. There are a number of experiments under way in the use of the IPv4 type of service or precedence bits to provide various forms of "differentiated service" for IP packets other than through the use of explicit flow setup. The traffic class field in the IPv6 header is intended to allow similar functionality to be supported in IPv6.

The expectation is that experimentation will eventually lead to agreement on what sorts of traffic classifications are most useful for IP packets. Detailed definitions of the syntax and semantics of all or some of the IPv6 Traffic Class bits, whether experimental or intended for eventual standardization, are to be provided in separate documents.

The following general requirements apply to the Traffic Class field:

■ The service interface to the IPv6 service within a node must provide a means for an upper-layer protocol to supply the value of the Traffic Class bits in packets originated by that upper-layer protocol. The default value must be zero for all eight bits.

■ Nodes that support a specific (experimental or eventual standard) use of some or all of the traffic class bits are permitted to change the value of those bits in packets that they originate, forward, or receive, as required for that specific use. Nodes should ignore and leave unchanged any bits of the Traffic Class field for which they do not support a specific use.

■ An upper-layer protocol must not assume that the value of the Traffic Class bits in a received packet are the same as the value sent by the packet's source.

5.8 Upper-Layer Protocol Issues

5.8.1 Upper-Layer Checksums

Any transport or other upper-layer protocol that includes the addresses from the IP header in its checksum computation must be modified for use over IPv6 to include the 128-bit IPv6 addresses instead of 32-bit IPv4 addresses. In particular, Figure 5.12 shows the TCP and UDP pseudoheader for IPv6:

■ If the IPv6 packet contains a Routing header, the destination address used in the pseudoheader is that of the final destination. At the originating node, that address will be in the last element of the Routing header; at the recipients, that address will be in the Destination Address field of the IPv6 header.

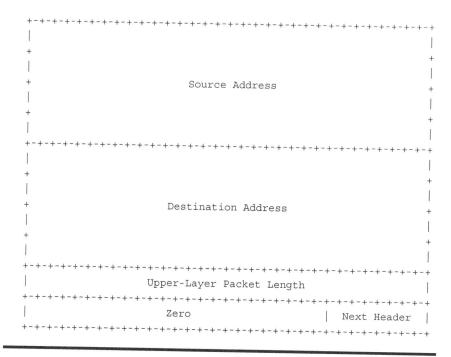

Figure 5.12   TCP and UDP pseudoheader for IPv6.

■ The Next Header value in the pseudoheader identifies the upper-layer proto-col (e.g., 6 for TCP or 17 for UDP). It will differ from the Next Header value in the IPv6 header if there are extension headers between the IPv6 header and the upper-layer header.

■ The upper-layer packet length in the pseudoheader is the length of the upper-layer header and data (e.g., TCP header plus TCP data). Some upper-layer protocols carry their own length information (e.g., the length field in the UDP header); for such protocols, this is the length used in the pseudoheader. Other protocols (such as TCP) do not carry their own length information, in which case the length used in the pseudoheader is the payload length from the IPv6 header minus the length of any extension headers present between the IPv6 header and the Upper-layer header.

■ Unlike IPv4, when UDP packets are originated by an IPv6 node, the UDP checksum is not optional. That is, whenever originating a UDP packet, an IPv6 node must compute a UDP checksum over the packet and the pseudo-header, and if that computation yields a result of zero, it must be changed to hex FFFF for placement in the UDP header. IPv6 receivers must discard UDP packets containing a zero checksum and should log the error.

The IPv6 version of ICMP includes the above pseudoheader in its checksum computation; this is a change from the IPv4 version of ICMP, which does not include a pseudoheader in its checksum. The reason for the change is to protect

ICMP from misdelivery or corruption of those fields of the IPv6 header on which it depends, which, unlike IPv4, are not covered by an IP layer checksum. The next header field in the pseudoheader for ICMP contains the value 58, which identifies the IPv6 version of ICMP.

5.8.2 Maximum Packet Lifetime

Unlike IPv4, IPv6 nodes are not required to enforce maximum packet lifetime; that is the reason the IPv4 Time To Live field was renamed Hop Limit in IPv6. In practice, very few, if any, IPv4 implementations conform to the requirement that they limit packet lifetime, so this is not a change in practice. Any upper-layer protocol that relies on the Internet layer (whether IPv4 or IPv6) to limit packet lifetime ought to be upgraded to provide its own mechanisms for detecting and discarding obsolete packets.

5.8.3 Maximum Upper-Layer Payload Size

When computing the maximum payload size available for upper-layer data, an upper-layer protocol must take into account the larger size of the IPv6 header relative to the IPv4 header. For example, in IPv4, TCP's maximum segment size (MSS) option is computed as the maximum packet size (a default value or a value learned through path MTU discovery) minus 40 octets (20 octets for the minimum-length IPv4 header and 20 octets for the minimum-length TCP header). When using TCP over IPv6, the MSS must be computed as the maximum packet size minus 60 octets because the minimum-length IPv6 header (i.e., an IPv6 header with no extension headers) is 20 octets longer than a minimum-length IPv4 header.

5.8.4 Responding to Packets Carrying Routing Headers

When an upper-layer protocol sends one or more packets in response to a received packet that included a Routing header, the response packets must not include a Routing header that was automatically derived by "reversing" the received Routing header unless the integrity and authenticity of the received Source Address and Routing header have been verified (e.g., via the use of an Authentication header in the received packet). In other words, only the following kinds of packets are permitted in response to a received packet bearing a Routing header:

- Response packets that do not carry Routing headers
- Response packets that carry Routing headers that were not derived by reversing the Routing header of the received packet (e.g., a Routing header supplied by local configuration)

■ Response packets that carry Routing headers that were derived by reversing the Routing header of the received packet if and only if the integrity and authenticity of the Source Address and Routing header from the received packet have been verified by the responder

5.9 Semantics and Usage of the Flow Label Field

A *flow* is a sequence of packets sent from a particular source to a particular (unicast or multicast) destination for which the source desires special handling by the intervening routers. The nature of that special handling might be conveyed to the routers by a control protocol, such as a resource reservation protocol, or by information within the flow's packets themselves (e.g., in a Hop-by-Hop Option).

There may be multiple active flows from a source to a destination, as well as traffic that is not associated with any flow. A flow is uniquely identified by the combination of a source address and a nonzero Flow Label. Packets that do not belong to a flow carry a Flow Label of zero.

A Flow Label is assigned to a flow by the flow's source node. New Flow Labels must be chosen (pseudo)randomly and uniformly from the range 1 to FFFFF hex. The purpose of the random allocation is to make any set of bits within the Flow Label field suitable for use as a hash key by routers for looking up the state associated with the flow.

All packets belonging to the same flow must be sent with the same Source Address, Destination Address, and Flow Label. If any of those packets includes a Hop-by-Hop Options header, then they all must be originated with the same Hop-by-Hop Options header contents (excluding the Next Header field of the Hop-by-Hop Options header). If any of those packets includes a Routing header, then they all must be originated with the same contents in all extension headers up to and including the Routing header (excluding the Next Header field in the Routing header). The routers or destinations are permitted, but not required, to verify that these conditions are satisfied. If a violation is detected, it should be reported to the source by an ICMP parameter problem message, Code 0, pointing to the high-order octet of the Flow Label field (i.e., offset one within the IPv6 packet).

The maximum lifetime of any flow-handling state established along a flow's path must be specified as part of the description of the state-establishment mechanism (e.g., the resource reservation protocol or the flow setup Hop-by-Hop Option). A source must not reuse a flow label for a new flow within the maximum lifetime of any flow-handling state that might have been established for the prior use of that flow label.

When a node stops and restarts (e.g., as a result of a "crash"), it must be careful not to use a flow label that it might have used for an earlier flow with a lifetime that may not have expired yet. This may be accomplished by recording flow label usage on stable storage so that it can be remembered across crashes or by refraining from using any flow labels until the maximum lifetime of any possible previously established flows has expired. If the minimum time for rebooting the node is known, that time can be deducted from the necessary waiting period before starting to allocate flow labels.

There is no requirement that all, or even most, packets belong to flows (i.e., carry nonzero Flow Labels). This observation is placed here to remind protocol designers and implementors not to assume otherwise. For example, it would be unwise to design a router with performance that would be adequate only if most packets belonged to flows or to design a header compression scheme that only worked on packets that belonged to flows.

5.10 Formatting Guidelines for Options

This section addresses how to lay out the fields when designing new options to be used in the Hop-by-Hop Options header or the Destination Options header, as described in Section 5.5.2. These guidelines are based on the following assumptions:

- One desirable feature is that any multi-octet fields within the option data area of an option be aligned on their natural boundaries; that is, fields of width n octets should be placed at an integer multiple of n octets from the start of the Hop-by-Hop or Destination Options header, for n = 1, 2, 4, or 8.
- Another desirable feature is that the Hop-by-Hop or Destination Options header take up as little space as possible, subject to the requirement that the header be an integer multiple of eight octets long.
- It may be assumed that, when either of the option-bearing headers are present, they carry a very small number of options, usually only one.

These assumptions suggest the following approach to laying out the fields of an option: Order the fields from smallest to largest, with no interior padding, then derive the alignment requirement for the entire option based on the alignment requirement of the largest field (up to a maximum alignment of eight octets). This approach is illustrated in the following examples:

Example 1

If an option X required two data fields, one of length 8 octets and one of length 4 octets, it would be laid out as follows:

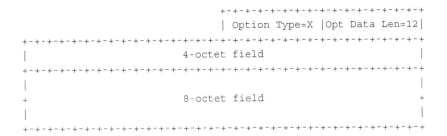

```
                              +-+-+-+-+-+-+-+-+-+-+-+-+-+-+-+-+
                              | Option Type=X |Opt Data Len=12|
   +-+-+-+-+-+-+-+-+-+-+-+-+-+-+-+-+-+-+-+-+-+-+-+-+-+-+-+-+-+-+-+-+
   |                         4-octet field                        |
   +-+-+-+-+-+-+-+-+-+-+-+-+-+-+-+-+-+-+-+-+-+-+-+-+-+-+-+-+-+-+-+-+
   |                                                              |
   +                        8-octet field                         +
   |                                                              |
   +-+-+-+-+-+-+-+-+-+-+-+-+-+-+-+-+-+-+-+-+-+-+-+-+-+-+-+-+-+-+-+-+
```

Its alignment requirement is 8n+2 to ensure that the eight-octet field starts at a multiple-of-eight offset from the start of the enclosing header. A complete Hop-by-Hop or Destination Options header containing this one option would look as follows:

```
+-+-+-+-+-+-+-+-+-+-+-+-+-+-+-+-+-+-+-+-+-+-+-+-+-+-+-+-+-+-+-+-+
|  Next Header  | Hdr Ext Len=1 | Option Type=X |Opt Data Len=12|
+-+-+-+-+-+-+-+-+-+-+-+-+-+-+-+-+-+-+-+-+-+-+-+-+-+-+-+-+-+-+-+-+
|                         4-octet field                         |
+-+-+-+-+-+-+-+-+-+-+-+-+-+-+-+-+-+-+-+-+-+-+-+-+-+-+-+-+-+-+-+-+
|                                                               |
+                         8-octet field                         +
|                                                               |
+-+-+-+-+-+-+-+-+-+-+-+-+-+-+-+-+-+-+-+-+-+-+-+-+-+-+-+-+-+-+-+-+
```

Example 2

If an option Y required three data fields, one of length four octets, one of length two octets, and one of length one octet, it would be laid out as follows:

```
                                                +-+-+-+-+-+-+-+-+
                                                | Option Type=Y |
+-+-+-+-+-+-+-+-+-+-+-+-+-+-+-+-+-+-+-+-+-+-+-+-+-+-+-+-+-+-+-+-+
|Opt Data Len=7 | 1-octet field |          2-octet field        |
+-+-+-+-+-+-+-+-+-+-+-+-+-+-+-+-+-+-+-+-+-+-+-+-+-+-+-+-+-+-+-+-+
|                         4-octet field                         |
+-+-+-+-+-+-+-+-+-+-+-+-+-+-+-+-+-+-+-+-+-+-+-+-+-+-+-+-+-+-+-+-+
```

Its alignment requirement is 4n+3 to ensure that the four-octet field starts at a multiple-of-four offset from the start of the enclosing header. A complete Hop-by-Hop or Destination Options header containing this one option would look as follows:

```
+-+-+-+-+-+-+-+-+-+-+-+-+-+-+-+-+-+-+-+-+-+-+-+-+-+-+-+-+-+-+-+-+
|  Next Header  | Hdr Ext Len=1 | Pad1 Option=0 | Option Type=Y |
+-+-+-+-+-+-+-+-+-+-+-+-+-+-+-+-+-+-+-+-+-+-+-+-+-+-+-+-+-+-+-+-+
|Opt Data Len=7 | 1-octet field |          2-octet field        |
+-+-+-+-+-+-+-+-+-+-+-+-+-+-+-+-+-+-+-+-+-+-+-+-+-+-+-+-+-+-+-+-+
|                         4-octet field                         |
+-+-+-+-+-+-+-+-+-+-+-+-+-+-+-+-+-+-+-+-+-+-+-+-+-+-+-+-+-+-+-+-+
| PadN Option=1 |Opt Data Len=2 |       0       |       0       |
+-+-+-+-+-+-+-+-+-+-+-+-+-+-+-+-+-+-+-+-+-+-+-+-+-+-+-+-+-+-+-+-+
```

Example 3

A Hop-by-Hop or Destination Options header containing both options X and Y from Examples 1 and 2 would have one of the two following formats, depending on which option appeared first:

```
+-+-+-+-+-+-+-+-+-+-+-+-+-+-+-+-+-+-+-+-+-+-+-+-+-+-+-+-+-+-+-+-+
| Next Header  | Hdr Ext Len=3 | Option Type=X |Opt Data Len=12|
+-+-+-+-+-+-+-+-+-+-+-+-+-+-+-+-+-+-+-+-+-+-+-+-+-+-+-+-+-+-+-+-+
|                          4-octet field                       |
+-+-+-+-+-+-+-+-+-+-+-+-+-+-+-+-+-+-+-+-+-+-+-+-+-+-+-+-+-+-+-+-+
|                                                              |
+                          8-octet field                       +
|                                                              |
+-+-+-+-+-+-+-+-+-+-+-+-+-+-+-+-+-+-+-+-+-+-+-+-+-+-+-+-+-+-+-+-+
| PadN Option=1 |Opt Data Len=1 |       0       | Option Type=Y |
+-+-+-+-+-+-+-+-+-+-+-+-+-+-+-+-+-+-+-+-+-+-+-+-+-+-+-+-+-+-+-+-+
|Opt Data Len=7 | 1-octet field |       2-octet field          |
+-+-+-+-+-+-+-+-+-+-+-+-+-+-+-+-+-+-+-+-+-+-+-+-+-+-+-+-+-+-+-+-+
|                          4-octet field                       |
+-+-+-+-+-+-+-+-+-+-+-+-+-+-+-+-+-+-+-+-+-+-+-+-+-+-+-+-+-+-+-+-+
| PadN Option=1 |Opt Data Len=2 |       0       |       0       |


+-+-+-+-+-+-+-+-+-+-+-+-+-+-+-+-+-+-+-+-+-+-+-+-+-+-+-+-+-+-+-+-+
+-+-+-+-+-+-+-+-+-+-+-+-+-+-+-+-+-+-+-+-+-+-+-+-+-+-+-+-+-+-+-+-+
| Next Header  | Hdr Ext Len=3 | Pad1 Option=0 | Option Type=Y |
+-+-+-+-+-+-+-+-+-+-+-+-+-+-+-+-+-+-+-+-+-+-+-+-+-+-+-+-+-+-+-+-+
|Opt Data Len=7 | 1-octet field |       2-octet field          |
+-+-+-+-+-+-+-+-+-+-+-+-+-+-+-+-+-+-+-+-+-+-+-+-+-+-+-+-+-+-+-+-+
|                          4-octet field                       |
+-+-+-+-+-+-+-+-+-+-+-+-+-+-+-+-+-+-+-+-+-+-+-+-+-+-+-+-+-+-+-+-+
| PadN Option=1 |Opt Data Len=4 |       0       |       0       |
+-+-+-+-+-+-+-+-+-+-+-+-+-+-+-+-+-+-+-+-+-+-+-+-+-+-+-+-+-+-+-+-+
|       0       |       0       | Option Type=X |Opt Data Len=12|
+-+-+-+-+-+-+-+-+-+-+-+-+-+-+-+-+-+-+-+-+-+-+-+-+-+-+-+-+-+-+-+-+
|                          4-octet field                       |
+-+-+-+-+-+-+-+-+-+-+-+-+-+-+-+-+-+-+-+-+-+-+-+-+-+-+-+-+-+-+-+-+
|                                                              |
+                          8-octet field                       +
|                                                              |
+-+-+-+-+-+-+-+-+-+-+-+-+-+-+-+-+-+-+-+-+-+-+-+-+-+-+-+-+-+-+-+-+
```

5.11 Introduction to Addressing

This section defines the addressing architecture of the IPv6 protocol [RFC4291]. The RFC includes the basic formats for the various types of IPv6 addresses (unicast, anycast, and multicast). This section covers the IPv6 addressing model, text representations of IPv6 addresses, and the definition of IPv6 unicast addresses, anycast addresses, and multicast addresses based on [RFC4291]. The discussion is for pedagogical purposes, and developers should refer to the latest IETF documentation.

5.11.1   IPv6 Addressing

IPv6 addresses are 128-bit identifiers for interfaces and sets of interfaces (with *interface* as defined in Section 5.3). There are three types of addresses:

- Unicast: An identifier for a single interface. A packet sent to a unicast address is delivered to the interface identified by that address.
- Anycast: An identifier for a set of interfaces (typically belonging to different nodes). A packet sent to an anycast address is delivered to one of the interfaces identified by that address (the "nearest" one, according to the routing protocols' measure of distance).
- Multicast: An identifier for a set of interfaces (typically belonging to different nodes). A packet sent to a multicast address is delivered to all interfaces identified by that address.

There are no broadcast addresses in IPv6 as their function is superseded by multicast addresses.

In [RFC4291], fields in addresses are given a specific name, for example *subnet*. When this name is used with the term *ID* for identifier after the name (e.g., subnet ID), it refers to the contents of the named field. When it is used with the term *prefix* (e.g., subnet prefix), it refers to all of the address from the left up to and including this field. In IPv6, all zeros and all ones are legal values for any field unless specifically excluded. Specifically, prefixes may contain, or end with, zero-value fields.

5.11.2   Addressing Model

IPv6 addresses of all types are assigned to interfaces, not nodes. An IPv6 unicast address refers to a single interface. Because each interface belongs to a single node, any of that node's interfaces' unicast addresses may be used as an identifier for the node.

All interfaces are required to have at least one link-local unicast address (see Section 5.11.8 for additional required addresses). A single interface may also have multiple IPv6 addresses of any type (unicast, anycast, and multicast) or scope. Unicast addresses with scope greater than link scope are not needed for interfaces that are not used as the origin or destination of any IPv6 packets to or from nonneighbors. This is sometimes convenient for point-to-point interfaces. There is one exception to this addressing model:

> A unicast address or a set of unicast addresses may be assigned to multiple physical interfaces if the implementation treats the multiple physical interfaces as one interface when presenting it to the internet layer. This is useful for load-sharing over multiple physical interfaces.

Currently, IPv6 continues the IPv4 model that a subnet prefix is associated with one link. Multiple subnet prefixes may be assigned to the same link.

5.11.3   Text Representation of Addresses

There are three conventional forms for representing IPv6 addresses as text strings:

1. The preferred form is x:x:x:x:x:x:x:x, with x indicating the hexadecimal values of the eight 16-bit pieces of the address, for example:

   ```
   FEDC:BA98:7654:3210:FEDC:BA98:7654:3210
   1080:0:0:0:8:800:200C:417A
   ```

   Note that it is not necessary to write the leading zeros in an individual field, but there must be at least one numeral in every field (except for the case described in Item 2).

2. Due to some methods of allocating certain styles of IPv6 addresses, it will be common for addresses to contain long strings of zero bits. To make writing addresses containing zero bits easier, a special syntax is available to compress the zeros. The use of :: indicates one or more groups of 16 bits of zeros. The :: can only appear once in an address. The :: can also be used to compress leading or trailing zeros in an address. For example, the following addresses:

   ```
   1080:0:0:0:8:800:200C:417A    a unicast address
   FF01:0:0:0:0:0:0:101          a multicast address
   0:0:0:0:0:0:0:1               the loopback address
   0:0:0:0:0:0:0:0               the unspecified addresses
   ```

   may be represented as:

   ```
   1080::8:800:200C:417A         a unicast address
   FF01::101                     a multicast address
   ::1                           the loopback address
   ::                            the unspecified addresses
   ```

3. An alternative form that is sometimes more convenient when dealing with a mixed environment of IPv4 and IPv6 nodes is x:x:x:x:x:x:d.d.d.d, with x indicating the hexadecimal values of the six high-order 16-bit pieces of the address and d indicating the decimal values of the four low-order 8-bit pieces of the address (standard IPv4 representation). For example:

   ```
   0:0:0:0:0:0:13.1.68.3
   0:0:0:0:0:FFFF:129.144.52.38
   ```

   or in compressed form:

```
::13.1.68.3
::FFFF:129.144.52.38
```

5.11.4 Text Representation of Address Prefixes

The text representation of IPv6 address prefixes is similar to the way IPv4 address prefixes are written in classless interdomain routing (CIDR) notation. An IPv6 address prefix is represented by the following notation:

```
ipv6-address/prefix-length
```

where

The ipv6-address is an IPv6 address in any of the notations listed in Section 5.11.3.

The prefix-length is a decimal value specifying how many of the leftmost contiguous bits of the address comprise the prefix.

For example, the following are legal representations of the 60-bit prefix 12AB00000000CD3 (hexadecimal):

```
12AB:0000:0000:CD30:0000:0000:0000:0000/60
12AB::CD30:0:0:0:0/60
12AB:0:0:CD30::/60
```

The following are not legal representations of the above prefix:

```
12AB:0:0:CD3/60    may drop leading zeros, but not trailing zeros,
                   within any 16-bit chunk of the address

12AB::CD30/60      address to left of "/" expands to
                   12AB:0000:0000:0000:0000:000:0000:CD30

12AB::CD3/60       address to left of "/" expands to
                   12AB:0000:0000:0000:0000:000:0000:0CD3
```

When writing both a node address and a prefix of that node address (e.g., the node's subnet prefix), the two can be combined as follows:

```
the node address     12AB:0:0:CD30:123:4567:89AB:CDEF
and its subnet number 12AB:0:0:CD30::/60
```

can be abbreviated as 12AB:0:0:CD30:123:4567:89AB:CDEF/60

5.11.5 Address Type Identification

The type of an IPv6 address is identified by the high-order bits of the address, as follows:

```
Address type            Binary prefix        IPv6 notation    Section
-----------             -------------        -------------    -------
Unspecified             00...0  (128 bits)   ::/128           5.12.5.2
Loopback                00...1  (128 bits)   ::1/128          5.12.5.3
Multicast               11111111             FF00::/8         5.12.7
Link-local unicast      1111111010           FE80::/10        5.12.5.6
Site-local unicast      1111111011           FEC0::/10        5.12.5.6
Global unicast          (everything else)
```

Anycast addresses are taken from the unicast address spaces (of any scope) and are not syntactically distinguishable from unicast addresses.

The general format of global unicast addresses is described in Section 5.11.6.4. Some special-purpose subtypes of global unicast addresses that contain embedded IPv4 addresses (for the purposes of IPv4-IPv6 interoperation) are described in Section 5.11.6.5.

Future specifications may redefine one or more subranges of the global unicast space for other purposes, but unless and until that happens, implementations must treat all addresses that do not start with any of the above-listed prefixes as global unicast addresses.

5.11.6 Unicast Addresses

IPv6 unicast addresses are aggregable with prefixes of arbitrary bit length similar to IPv4 addresses under CIDR.

There are several types of unicast addresses in IPv6, in particular global unicast, site-local unicast, and link-local unicast. There are also some special-purpose subtypes of global unicast, such as IPv6 addresses with embedded IPv4 addresses or encoded network service access point (NSAP) addresses. Additional address types or subtypes can be defined in the future.

IPv6 nodes may have considerable or little knowledge of the internal structure of the IPv6 address, depending on the role the node plays (i.e., host versus router). At a minimum, a node may consider that unicast addresses (including its own) have no internal structure:

```
|                           128 bits                            |
+---------------------------------------------------------------+
|                         node address                          |
+---------------------------------------------------------------+
```

A slightly sophisticated host (but still rather simple) may also be aware of subnet prefixes for the links to which it is attached, for which different addresses may have different values for n:

```
|                      n bits                    |   128-n bits   |
+-----------------------------------------------+----------------+
|                   subnet prefix                |  interface ID  |
+-----------------------------------------------+----------------+
```

Although a very simple router may have no knowledge of the internal structure of IPv6 unicast addresses, routers will more generally have knowledge of one or more of the hierarchical boundaries for the operation of routing protocols. The known boundaries will differ from router to router, depending on what positions the router holds in the routing hierarchy.

5.11.6.1 Interface Identifiers

Interface identifiers in IPv6 unicast addresses are used to identify interfaces on a link. They are required to be unique within a subnet prefix. It is recommended that the same interface identifier not be assigned to different nodes on a link. They may also be unique over a broader scope. In some cases, an interface's identifier will be derived directly from that interface's link-layer address. The same interface identifier may be used on multiple interfaces on a single node as long as they are attached to different subnets.

Note that the uniqueness of interface identifiers is independent of the uniqueness of IPv6 addresses. For example, a global unicast address may be created with a nonglobal scope interface identifier and a site-local address may be created with a global scope interface identifier.

For all unicast addresses, except those that start with binary value 000, interface IDs are required to be 64 bits long and to be constructed in the modified 64-bit extended unique identifier (EUI-64) format.

Modified EUI-64 format based interface identifiers may have global scope when derived from a global token (e.g., Institute of Electrical and Electronics Engineers [IEEE] 802 48-bit Media Access Control (MAC) or IEEE EUI-64 identifiers) or may have local scope if a global token is not available (e.g., serial links, tunnel endpoints, etc.) or global tokens are undesirable (e.g., temporary tokens for privacy).

Modified EUI-64 format interface identifiers are formed by inverting the "u" bit (universal/local bit in IEEE EUI-64 terminology) when forming the interface identifier from IEEE EUI-64 identifiers. In the resulting modified EUI-64 format, the u bit is set to one (1) to indicate global scope, and it is set to zero (0) to indicate local scope. The first three octets in binary of an IEEE EUI-64 identifier are as follows:

```
 0          0 0        1 1        2
|0          7 8        5 6        3|
 +----+----+----+----+----+----+
|cccc|ccug|cccc|cccc|cccc|cccc|
 +----+----+----+----+----+----+
```

written in Internet standard bit order, where u is the universal/local bit, g is the individual/group bit, and c are the bits of the company_id. Section 5.13 provides examples of the creation of modified EUI-64 format-based interface identifiers.

The motivation for inverting the u bit when forming an interface identifier is to make it easy for system administrators to manually configure nonglobal identifiers when hardware tokens are not available. This is expected to be the case for serial links, tunnel endpoints, and so on. The alternative would have been for these to be of the form 0200:0:0:1, 0200:0:0:2, and so on, instead of the much simpler 1, 2, and so on.

The use of the universal/local bit in the modified EUI-64 format identifier is to allow development of future technology that can take advantage of interface identifiers with global scope.

5.11.6.2 The Unspecified Address

The address 0:0:0:0:0:0:0:0 is called the unspecified address. It must never be assigned to any node. It indicates the absence of an address. One example of its use is in the source address field of any IPv6 packets sent by an initializing host before it has learned its own address.

The unspecified address must not be used as the destination address of IPv6 packets or in IPv6 routing headers. An IPv6 packet with a source address of unspecified must never be forwarded by an IPv6 router.

5.11.6.3 The Loopback Address

The unicast address 0:0:0:0:0:0:0:1 is called the loopback address. It may be used by a node to send an IPv6 packet to itself. It may never be assigned to any physical interface. It is treated as having link-local scope and may be thought of as the link-local unicast address of a virtual interface (typically called the loopback interface) to an imaginary link that goes nowhere.

The loopback address must not be used as the source address in IPv6 packets that are sent outside of a single node. An IPv6 packet with a destination address of loopback must never be sent outside of a single node and must never be forwarded by an IPv6 router. A packet received on an interface with a destination address of loopback must be dropped.

5.11.6.4 Global Unicast Addresses

The general format for IPv6 global unicast addresses is as follows:

```
|          n bits          |   m bits  |      128-n-m bits        |
+--------------------------+-----------+--------------------------+
| global routing prefix    | subnet ID |      interface ID        |
+--------------------------+-----------+--------------------------+
```

where the global routing prefix is a (typically hierarchically structured) value assigned to a site (a cluster of subnets/links), the subnet ID is an identifier of a link within the site, and the interface ID is as defined in Section 5.11.6.1.

All global unicast addresses other than those that start with binary 000 have a 64-bit interface ID field (i.e., n + m = 64), formatted as described in Section 5.11.6.1. Global unicast addresses that start with binary 000 have no such constraint on the size or structure of the interface ID field.

Examples of global unicast addresses that start with binary 000 are the IPv6 address with embedded IPv4 addresses described in Section 5.11.6.5 and the IPv6 address containing encoded NSAP addresses.

5.11.6.5 IPv6 Addresses with Embedded IPv4 Addresses

The IPv6 transition mechanisms include a technique for hosts and routers to dynamically tunnel IPv6 packets over an IPv4 routing infrastructure. IPv6 nodes that use this technique are assigned special IPv6 unicast addresses that carry a global IPv4 address in the low-order 32 bits. This type of address is termed an IPv4-compatible IPv6 address and has the following format:

```
|                80 bits                | 16  |       32 bits      |
+---------------------------------------+-----+--------------------+
|0000.............................0000|0000|    IPv4 address    |
+---------------------------------------+----+--------------------+
```

Note that the IPv4 address used in the IPv4-compatible IPv6 address must be a globally unique IPv4 unicast address.

A second type of IPv6 address that holds an embedded IPv4 address is also defined. This address type is used to represent the addresses of IPv4 nodes as IPv6 addresses. This type of address is termed an IPv4-mapped IPv6 address and has the following format:

```
|                80 bits                | 16  |       32 bits      |
+---------------------------------------+-----+--------------------+
|0000............................0000|FFFF|    IPv4 address    |
+---------------------------------------+----+--------------------+
```

5.11.6.6 Local-Use IPv6 Unicast Addresses

There are two types of local-use unicast addresses defined. These are link-local and site-local. The link-local is for use on a single link, and the site-local is for use in a single site. Link-local addresses have the following format:

```
|   10     |
|  bits    |          54 bits          |            64 bits            |
+----------+---------------------------+-------------------------------+
|1111111010|             0             |          interface ID         |
+----------+---------------------------+-------------------------------+
```

Link-local addresses are designed to be used for addressing on a single link for purposes such as automatic address configuration, neighbor discovery, or when no routers are present.

Routers must not forward any packets with link-local source or destination addresses to other links.

Site-local addresses have the following format:

```
|   10     |
|  bits    |          54 bits          |            64 bits            |
+----------+---------------------------+-------------------------------+
|1111111011|         subnet ID         |          interface ID         |
+----------+---------------------------+-------------------------------+
```

Site-local addresses are designed to be used for addressing inside of a site without the need for a global prefix. Although a subnet ID may be up to 54 bits long, it is expected that globally connected sites will use the same subnet IDs for site-local and global prefixes.

Routers must not forward any packets with site-local source or destination addresses outside of the site.

5.11.7 Anycast Addresses

An IPv6 anycast address is an address that is assigned to more than one interface (typically belonging to different nodes), with the property that a packet sent to an anycast address is routed to the "nearest" interface having that address, according to the routing protocols' measure of distance.

Anycast addresses are allocated from the unicast address space using any of the defined unicast address formats. Thus, anycast addresses are syntactically indistinguishable from unicast addresses. When a unicast address is assigned to more than one interface, thus turning it into an anycast address, the nodes to which the address is assigned must be explicitly configured to know that it is an anycast address.

For any assigned anycast address, there is a longest prefix P of that address that identifies the topological region in which all interfaces belonging to that anycast address reside. Within the region identified by P, the anycast address must be maintained as a separate entry in the routing system (commonly referred to as a host route); outside the region identified by P, the anycast address may be aggregated into the routing entry for prefix P.

Note that, in the worst case, the prefix P of an anycast set may be the null prefix; that is, the members of the set may have no topological locality. In that case, the anycast address must be maintained as a separate routing entry throughout the entire Internet, which presents a severe scaling limit on how many such "global" anycast sets may be supported. Therefore, it is expected that support for global anycast sets may be unavailable or very restricted.

One expected use of anycast addresses is to identify the set of routers belonging to an organization providing Internet service. Such addresses could be used as intermediate addresses in an IPv6 routing header to cause a packet to be delivered via a particular service provider or sequence of service providers.

Some other possible uses are to identify the set of routers attached to a particular subnet or the set of routers providing entry into a particular routing domain.

There is little experience with widespread, arbitrary use of Internet anycast addresses and some known complications and hazards when using them in their full generality. Until more experience has been gained and solutions are specified, the following restrictions are imposed on IPv6 anycast addresses:

- An anycast address must not be used as the source address of an IPv6 packet.
- An anycast address must not be assigned to an IPv6 host; that is, it may be assigned to an IPv6 router only.

5.11.7.1 Required Anycast Address

The subnet-router anycast address is predefined. Its format is as follows:

```
|                       n bits                      |  128-n bits    |
+--------------------------------------------------+----------------+
|                    subnet prefix                 | 00000000000000 |
+--------------------------------------------------+----------------+
```

The subnet prefix in an anycast address is the prefix that identifies a specific link. This anycast address is syntactically the same as a unicast address for an interface on the link with the interface identifier set to zero.

Packets sent to the subnet-router anycast address will be delivered to one router on the subnet. All routers are required to support the subnet-router anycast addresses for the subnets to which they have interfaces.

The subnet-router anycast address is intended to be used for applications in which a node needs to communicate with any one of the set of routers.

5.11.8 Multicast Addresses

An IPv6 multicast address is an identifier for a group of interfaces (typically on different nodes). An interface may belong to any number of multicast groups. Multicast addresses have the following format:

```
|   8    | 4 | 4 |                    112 bits                   |
+------- -+----+----+---------------------------------------------+
|11111111|Flgs|Scop|               Group ID                      |
+--------+----+----+---------------------------------------------+
```

- Binary 11111111 at the start of the address identifies the address as being a multicast address.
- Flgs is a set of 4 flags: |0|0|0|T|
  The high-order three flags are reserved and must be initialized to 0.
  T = 0 indicates a permanently assigned (well-known) multicast address, assigned by the Internet Assigned Number Authority (IANA).
  T = 1 indicates a nonpermanently assigned (transient) multicast address.
- Scop is a 4-bit multicast scope value used to limit the scope of the multicast group. The values are as follows:

  0 reserved
  1 interface-local scope
  2 link-local scope
  3 reserved
  4 admin-local scope
  5 site-local scope
  6 (unassigned)
  7 (unassigned)
  8 organization-local scope
  9 (unassigned)
  A (unassigned)
  B (unassigned)
  C (unassigned)
  D (unassigned)
  E global scope
  F reserved

  Interface-local scope spans only a single interface on a node and is useful only for loopback transmission of multicast.

Link-local and site-local multicast scopes span the same topological regions as the corresponding unicast scopes.

Admin-local scope is the smallest scope that must be administratively configured (i.e., not automatically derived from physical connectivity or other, non-multicast-related configuration).

Organization-local scope is intended to span multiple sites belonging to a single organization.

Scopes labeled (unassigned) are available for administrators to define additional multicast regions.

■ Group ID identifies the multicast group, either permanent or transient, within the given scope.

The "meaning" of a permanently assigned multicast address is independent of the scope value. For example, if the Network Time Protocol (NTP) servers group is assigned a permanent multicast address with a Group ID of 101 (hex), then:

FF01:0:0:0:0:0:0:101 means all NTP servers on the same interface (i.e., the same node) as the sender.
FF02:0:0:0:0:0:0:101 means all NTP servers on the same link as the sender.
FF05:0:0:0:0:0:0:101 means all NTP servers in the same site as the sender.
FF0E:0:0:0:0:0:0:101 means all NTP servers in the Internet.

Nonpermanently assigned multicast addresses are meaningful only within a given scope. For example, a group identified by the nonpermanent, site-local multicast address FF15:0:0:0:0:0:0:101 at one site bears no relationship to a group using the same address at a different site to a nonpermanent group using the same Group ID with different scope, or to a permanent group with the same Group ID.

Multicast addresses must not be used as source addresses in IPv6 packets or appear in any routing header.

Routers must not forward any multicast packets beyond the scope indicated by the scop field in the destination multicast address.

Nodes must not originate a packet to a multicast address with a Scop field that contains the reserved value 0; if such a packet is received, it must be silently dropped. Nodes should not originate a packet to a multicast address with a scop field that contains the reserved value F; if such a packet is sent or received, it must be treated the same as packets destined to a global (Scop E) multicast address.

5.11.8.1 Predefined Multicast Addresses

The following well-known multicast addresses are predefined. The Group IDs defined in this section are defined for explicit scope values. Use of these Group IDs for any other scope values, with the T flag equal to zero, is not allowed.

```
Reserved Multicast Addresses:    FF00:0:0:0:0:0:0:0
                                 FF01:0:0:0:0:0:0:0
                                 FF02:0:0:0:0:0:0:0
                                 FF03:0:0:0:0:0:0:0
                                 FF04:0:0:0:0:0:0:0
                                 FF05:0:0:0:0:0:0:0
                                 FF06:0:0:0:0:0:0:0
                                 FF07:0:0:0:0:0:0:0
                                 FF08:0:0:0:0:0:0:0
                                 FF09:0:0:0:0:0:0:0
                                 FF0A:0:0:0:0:0:0:0
                                 FF0B:0:0:0:0:0:0:0
                                 FF0C:0:0:0:0:0:0:0
                                 FF0D:0:0:0:0:0:0:0
                                 FF0E:0:0:0:0:0:0:0
                                 FF0F:0:0:0:0:0:0:0
```

The above multicast addresses are reserved and shall never be assigned to any multicast group.

```
All Nodes Addresses:    FF01:0:0:0:0:0:0:1
                        FF02:0:0:0:0:0:0:1
```

The above multicast addresses identify the group of all IPv6 nodes, within scope one (interface local) or two (link-local).

```
All Routers Addresses:    FF01:0:0:0:0:0:0:2
                          FF02:0:0:0:0:0:0:2
                          FF05:0:0:0:0:0:0:2
```

The above multicast addresses identify the group of all IPv6 routers within scope one (interface local), two (link local), or five (site local).

```
Solicited-Node Address:    FF02:0:0:0:0:1:FFXX:XXXX
```

Solicited-node multicast addresses are computed as a function of a node's unicast and anycast addresses. A solicited-node multicast address is formed by taking the low-order 24 bits of an address (unicast or anycast) and appending those bits to the prefix FF02:0:0:0:0:1:FF00::/104 resulting in a multicast address in the range

```
FF02:0:0:0:0:1:FF00:0000
```

to

```
FF02:0:0:0:0:1:FFFF:FFFF
```

For example, the solicited node multicast address corresponding to the IPv6 address `4037::01:800:200E:8C6C` is `FF02::1:FF0E:8C6C`. IPv6 addresses that differ only in the high-order bits (e.g., due to multiple high-order prefixes associated with different aggregations) will map to the same solicited-node address, thereby reducing the number of multicast addresses a node must join.

A node is required to compute and join (on the appropriate interface) the associated solicited-node multicast addresses for every unicast and anycast address it is assigned.

5.11.9 A Node's Required Addresses

A host is required to recognize the following addresses as identifying itself:

- Its required link-local address for each interface.
- Any additional unicast and anycast addresses that have been configured for the node's interfaces (manually or automatically).
- The loopback address.
- The all-nodes multicast addresses defined in Section 5.11.8.1.
- The solicited-node multicast address for each of its unicast and anycast addresses.
- Multicast addresses of all other groups to which the node belongs.

A router is required to recognize all addresses that a host is required to recognize, plus the following addresses as identifying itself:

- The subnet-router anycast addresses for all interfaces for which it is configured to act as a router.
- All other anycast addresses with which the router has been configured.
- The all-routers multicast addresses defined in Section 5.11.8.1.

5.12 IANA Considerations

The initial assignment of IPv6 address space is as follows:

Allocation	Prefix (binary)	Fraction of Address Space
Unassigned (see Note 1 below)	0000 0000	1/256
Unassigned	0000 0001	1/256
Reserved for NSAP Allocation	0000 001	1/128 (RFC1888)
Unassigned	0000 01	1/64

```
Unassigned                       0000 1        1/32
Unassigned                       0001          1/16
Global Unicast                   001           1/8    (per [RFC2374])
Unassigned                       010           1/8
Unassigned                       011           1/8
Unassigned                       100           1/8
Unassigned                       101           1/8
Unassigned                       110           1/8
Unassigned                       1110          1/16
Unassigned                       1111 0        1/32
Unassigned                       1111 10       1/64
Unassigned                       1111 110      1/128
Unassigned                       1111 1110 0   1/512
Link-Local Unicast Addresses     1111 1110 10  1/1024
Site-Local Unicast Addresses     1111 1110 11  1/1024
Multicast Addresses              1111 1111     1/256
```

Note the following:

1. The unspecified address, the loopback address, and the IPv6 addresses with embedded IPv4 addresses are assigned out of the 0000 0000 binary prefix space.
2. For now, IANA should limit its allocation of IPv6 unicast address space to the range of addresses that start with binary value 001. The rest of the global unicast address space (approximately 85 percent of the IPv6 address space) is reserved for future definition and use and is not to be assigned by IANA at this time.

5.13 Creating Modified EUI-64 Format Interface Identifiers

Depending on the characteristics of a specific link or node, there are a number of approaches for creating modified EUI-64 format interface identifiers. This section describes some of these approaches.

EUI is defined in IEEE's Guidelines for 64-bit Global Identifier (EUI-64) Registration Authority, March 1997 (see Section 5.14).

5.13.1 *Links or Nodes with IEEE EUI-64 Identifiers*

The only change needed to transform an IEEE EUI-64 identifier to an interface identifier is to invert the u (universal/local) bit. For example, a globally unique IEEE EUI-64 identifier of the following form:

```
|0                 1|1                 3|3                 4|4                 6|
|0                 5|6                 1|2                 7|8                 3|
+----------------+----------------+----------------+----------------+
|cccccc0gcccccccc|ccccccccmmmmmmmm|mmmmmmmmmmmmmmmm|mmmmmmmmmmmmmmmm|
+----------------+----------------+----------------+----------------+
```

where c are the bits of the assigned company_id, 0 is the value of the universal/local bit to indicate global scope, g is the individual/group bit, and m are the bits of the manufacturer-selected extension identifier. The IPv6 interface identifier would be of the following form:

```
|0                 1|1                 3|3                 4|4                 6|
|0                 5|6                 1|2                 7|8                 3|
+----------------+----------------+----------------+----------------+
|cccccc1gcccccccc|ccccccccmmmmmmmm|mmmmmmmmmmmmmmmm|mmmmmmmmmmmmmmmm|
+----------------+----------------+----------------+----------------+
```

The only change is inverting the value of the universal/local bit.

5.13.2 Links or Nodes with IEEE 802 48-Bit MACs

EUI64 defines a method to create an IEEE EUI-64 identifier from an IEEE 48-bit MAC identifier. This is to insert two octets, with hexadecimal values of 0xFF and 0xFE, in the middle of the 48-bit MAC (between the company_id and vendor supplied ID). For example, the 48-bit IEEE MAC with global scope:

```
|0                 1|1                 3|3                 4|
|0                 5|6                 1|2                 7|
+----------------+----------------+----------------+
|cccccc0gcccccccc|ccccccccmmmmmmmm|mmmmmmmmmmmmmmmm|
+----------------+----------------+----------------+
```

where c are the bits of the assigned company_id, 0 is the value of the universal/local bit to indicate global scope, g is the individual/group bit, and m are the bits of the manufacturer-selected extension identifier. The interface identifier would be of the following form:

```
|0                 1|1                 3|3                 4|4                 6|
|0                 5|6                 1|2                 7|8                 3|
+----------------+----------------+----------------+----------------+
|cccccc1gcccccccc|cccccccc11111111|11111110mmmmmmmm|mmmmmmmmmmmmmmmm|
+----------------+----------------+----------------+----------------+
```

When IEEE 802 48-bit MAC addresses are available (on an interface or a node), an implementation may use them to create interface identifiers due to their availability and uniqueness properties.

5.13.3 *Links with Other Kinds of Identifiers*

There are a number of types of links that have link-layer interface identifiers other than IEEE EIU-64 or IEEE 802 48-bit MACs. Examples include LocalTalk and Arcnet. The method to create a modified EUI-64 format identifier is to take the link identifier (e.g., the LocalTalk eight-bit node identifier) and zero fill it to the left. For example, a LocalTalk eight-bit node identifier of hexadecimal value 0x4F results in the following interface identifier:

```
|0                1|1              3|3              4|4              6|
|0                5|6              1|2              7|8              3|
+----------------+----------------+----------------+----------------+
|0000000000000000|0000000000000000|0000000000000000|0000000001001111|
+----------------+----------------+----------------+----------------+
```

Note that this results in the universal/local bit set to zero to indicate local scope.

5.13.4 *Links without Identifiers*

There are a number of links that do not have any type of built-in identifier. The most common of these are serial links and configured tunnels. Interface identifiers must be chosen that are unique within a subnet prefix.

When no built-in identifier is available on a link, the preferred approach is to use a global interface identifier from another interface or one that is assigned to the node itself. When using this approach, no other interface connecting the same node to the same subnet prefix may use the same identifier.

If there is no global interface identifier available for use on the link, the implementation needs to create a local-scope interface identifier. The only requirement is that it be unique within a subnet prefix. There are many possible approaches to select a subnet-prefix-unique interface identifier. These include the following:

Manual configuration
Node serial number
Other Node-Specific Token

The subnet-prefix-unique interface identifier should be generated in such a manner that does not change after a reboot of a node or if interfaces are added or deleted from the node.

The selection of the appropriate algorithm is link and implementation dependent. It is strongly recommended that a collision detection algorithm be implemented as part of any automatic algorithm.

5.14   64-Bit Global Identifier (EUI-64) Registration Authority

The IEEE-defined EUI-64 is a concatenation of the 24-bit company_id value by the IEEE Registration Authority and a 40-bit extension identifier assigned by the organization with that company_id assignment. The IEEE administers the assignment of 24-bit *company_id* values. The assignments of these values are public, so that a user of an EUI-64 value can identify the manufacturer that provided any value. The IEEE Registration Authority Committee (IEEE/RAC) has no control over the assignments of 40-bit extension identifiers and assumes no liability for assignments of duplicate EUI-64 identifiers by manufacturers.

5.14.1   Application Restrictions

Given the minimal probability of consuming all the EUI-64 identifiers, the IEEE/RAC places minimal restrictions on their use within standards. However, if used within the context of an IEEE standard, the documentation shall be reviewed by the IEEE/RAC for correctness and clarity. The IEEE/RAC shall not otherwise restrict the use of EUI-64 identifiers within standards. If the EUI-64 is referenced within non-IEEE standards, there shall not be any reference to IEEE unless approved by the IEEE/RAC.

5.14.2   Distribution Restrictions

Given the minimal probability of consuming all the EUI-64 identifiers, the IEEE/RAC places minimal restrictions on their redistribution through third parties, as follows:

1. Allocation. The EUI-64 values shall be sold within electronically readable parts; no more than one EUI-64 value shall be contained within each component that is manufactured.
2. Packaging. A component containing the EUI-64 value shall have a distinguishing characteristic (such as color or shape) to distinguish it from other commonly used identifier components.
3. Documentation. Documentation should be readily available.
4. Legal indemnification. Any organization producing EUI-64 values shall indemnify the IEEE for damages arising from duplicate number assignments.

The term EUI-64 is trademarked by the IEEE. Companies are allowed to use this term for commercial purposes, but only if their use of this term has been reviewed by the IEEE/RAC and the proposed products using the EUI-64 conform to these restrictions.

5.14.3 Application Documentation

As a condition for receiving a *company_id* assignment, a manufacturer of EUI-64 values accepts the following responsibilities:

1. This documentation shall be readily available (at no cost) to any purchaser of EUI-64 values.
2. The manufacturer's part specification should include an unambiguous description of how the EUI-64 value is accessed (pin or address descriptions).

5.14.4 Manufacturer-Assigned Identifiers

The manufacturer identifier assignment allows the assignee to generate approximately 1 trillion (10^{12}) unique EUI-64 values by varying the last 40 bits. The IEEE intends not to assign another OUI/*company_id* value to a manufacturer of EUI-64 values until the manufacturer has consumed, in product, the preponderance (more than 90 percent) of this block of potential unique words. It is incumbent on the manufacturer to ensure that large portions of the unique word block are not left unused in manufacturing.

5.15 Additional Technical Details

As indicated in Chapter 1, the IPv6 protocol apparatus is described by the 100+ RFCs identified in Appendix B (some have been obsoleted or replaced). The interested reader, particularly developers, should work through that body of information for additional technical details.

References

[RFC791] Information Science Institute, Internet Protocol, RFC 791, September 1981.

[RFC1661] W. Simpson, The Point-to-Point Protocol, RFC 1661, July 1994.

[RFC1981] J. McCann, S. Deering, J. Mogul, Path MTU Discovery for IP version 6, RFC 1981, August 1996.

[RFC2374] R. Hinden, M. O'Dell, S. Deering, An IPv6 Aggregatable Global Unicast Address Format, RFC 2374, July 1998.

[RFC2402] S. Kent, R. Atkinson, IP anthentication Header, RFC 2402, November 1998.

[RFC2406] S. Kent, R. Atlinson, IP Encapsulating Security Payload, RFC 2406, November 1998.

[RFC2460] S. Deering and R. Hinden, Internet Protocol, Version 6 (IPv6) Specification, RFC 2460, December 1998.

[RFC2483] M. Mealling, R. Daniel Jr., URI Resolution Services, RFC 2483, January 1999.

[RFC3232] J. Reynolds, Assigned Numbers, RFC 3232, January 2002.

[RFC35958] B. Wijnen, Textual Conventions for IPv6 Flow Label, RFC 3595, September 2003.

[RFC3697] J. Rajahalme, A. Conta, B. Carpenter, S. Deering, IPv6 Flow Label Specification, RFC 3697, March 2004.

[RFC4291] R. Hinden and S. Deering, Internet Protocol Version 6 (IPv6) Addressing Architecture, RFC 4291, February 2006.

Chapter 6

Transition Approaches and Mechanisms

6.1 Introduction

Although most technical aspects of Internet Protocol version 6 (IPv6) have been defined for some time, deployment of IPv6 is occurring gradually. Initially, IPv6 is to be deployed within isolated islands with interconnectivity among the islands achieved by the existing IPv4 infrastructure; a number of transition mechanisms have been defined to interconnect such islands.

There is an additional need for support for IPv6 hosts and routers that need to interoperate with legacy IPv4 hosts; an overview of such mechanisms for this purpose is provided in [RFC2893]. That RFC defines the following types of nodes with respect to the transition to IPv6:

IPv4-only node: A host or router that implements only IPv4. An IPv4-only node does not understand IPv6. The installed base of IPv4 hosts and routers are examples of IPv4-only nodes.

IPv6/IPv4 node: A host or router that implements both the IPv4 and IPv6 protocols.

IPv6-only node: A host or router that implements IPv6 and does not implement IPv4.

IPv6 node: Any host or router that implements IPv6. IPv6/IPv4 and IPv6-only nodes are both IPv6 nodes.

IPv4 node: Any host or router that implements IPv4. IPv6/IPv4 and IPv4-only nodes are both IPv4 nodes.

The RFC also defines the IPv4-compatible IPv6 address (e.g., ::156.55.23.5), discussed previously in Section 5.11.6.5 on IPv6 address space. IPv4-compatible IPv6 addresses are used to implement a simple automatic tunneling mechanism discussed in 6.1.3.2.

In addition to connectivity issues at the IP layer, the transition to IPv6 is also not entirely transparent to the networking layers above IP. As discussed previously, IPv6 addresses are significantly longer in size than IPv4 addresses and thus will require a change in application programming interfaces (APIs) or service primitive parameters that include IP addresses. Applications must also be extended to select the appropriate protocol, IPv4 or IPv6, when a Domain Name System (DNS) lookup returns both types of addresses. In general, legacy applications written for IPv4 need to be either rewritten or amended to support IPv6. For example, the application layer File Transfer Protocol (FTP) embeds IP addresses in its protocol fields and could thus require changes to both the client and server FTP applications.

The Internet Engineering Task Force (IETF) has defined a number of specific mechanisms to assist in transitioning to IPv6. These mechanisms are generally classified as belonging to the following categories:

Dual Stack: The principal building block for transitioning is the dual-stack approach. Dual-stack nodes, as the name suggests, maintain two protocol stacks that operate in parallel and thus allow the end system or router to operate via either protocol. In end systems, they enable both IPv4- and IPv6-capable applications to operate on the same node. Dual-stack capabilities in routers allow handling of both IPv4 and IPv6 packet types.

Translation: Translation refers to the direct conversion of protocols (e.g., between IPv4 and IPv6) and may include transformation of both the protocol header and the protocol payload. Translation can occur at several layers in the protocol stack, including IP, transport, and application layers. Note that protocol translation can result in feature loss when there is no clear mapping between the features provided by translated protocols. For instance, translation of an IPv6 header into an IPv4 header will lead to the loss of the IPv6 flow label and its accompanying functionality.

Tunneling (or encapsulation): Tunneling is used to interconnect compatible networking nodes or domains across incompatible networks. It can be viewed technically as the transfer of a payload protocol data unit by an encapsulating carrier protocol. For IPv6 transition, the IPv6 protocol data unit is generally carried as the payload of an IPv4 packet. Encapsulation of the payload protocol data unit is performed at the tunnel entrance, and decapsulation is performed at the tunnel exit point.

Note that a transition mechanism may employ techniques from more than one of these categories. For example, when an end system or router creates an IPv6-in-IPv4 tunnel, this could be classified as both dual stack (having both an IPv4 and IPv6 address) and tunneling.

This chapter examines a number of transition mechanisms that have been standardized and widely supported and, for historical purposes, a few that have been deprecated.

6.2 IPv6/IPv4 Dual Stack

In the dual-stack scheme (RFC 2893), a network node incorporates both IPv4 and IPv6 protocol stacks in parallel (see Figure 6.1). IPv4 applications use the IPv4 stack, and IPv6 applications use the IPv6 stack. Flow decisions in the node are based on the IP header *version* field for packets that are received from the lower layers — a *version* field with a value of four results in passing the IP protocol data unit to the IPv4 layer and a value of six to the IPv6 layer. When sending packets, the destination address type received from the upper layers determines the appropriate stack. The address types typically come from DNS lookups; the appropriate stack is chosen in response to returned DNS record types.

Many off-the-shelf commercial operating systems already provide dual-IP protocol stacks. For example, Microsoft® Windows XP® and Windows Server® 2003 operating systems implement the dual-stack architecture shown in Figure 6.1. Windows Vista® operating system[1] implements what is termed a next-generation Transmission Control Protocol/Internet Protocol (TCP/IP) stack that incorporates the dual-stack architecture, but both IPv4 and IPv6 share common transport and framing layers, unlike the transport layers shown in Figure 6.1 [MIC200701]. Consequently, the dual-stack mechanism will be widely deployed as a transition mechanism. However, note that dual stacks only enable like applications to communicate (i.e., IPv6 to IPv6 and IPv4 to IPv4). The next section discusses mechanisms that would allow communications between IPv4-only legacy nodes and IPv6-only nodes.

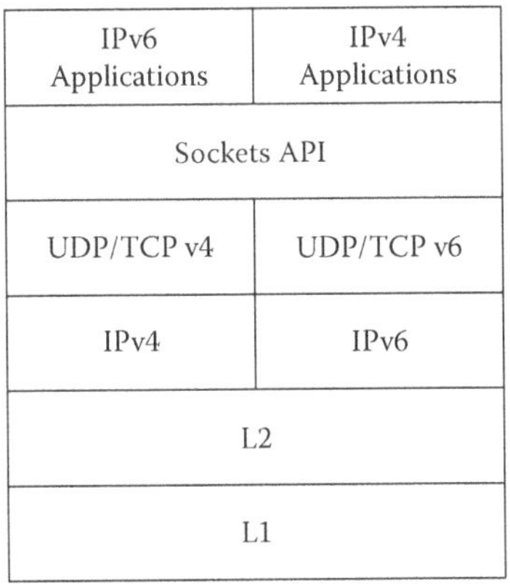

IPv6 Applications	IPv4 Applications
Sockets API	
UDP/TCP v4	UDP/TCP v6
IPv4	IPv6
L2	
L1	

Figure 6.1   End-system dual-stack transition mechanism.

---

[1] Microsoft, Windows, Windows Server, and Windows Vista are registered trademarks of Microsoft Corporation in the United States and other countries.

6.3 Translation Mechanisms

6.3.1 Stateless Internet Protocol/Internet Control Messaging Protocol Translation (SIIT)

6.3.1.1 Overview

The Stateless Internet Protocol/Internet Control Messaging Protocol Translation (SIIT) [RFC2765] is an IPv6 transition mechanism that allows IPv6-only hosts to talk to IPv4-only hosts. The mechanism involves a stateless mapping or bidirectional translation algorithm between IPv4 and IPv6 packet headers as well as between Internet Control Messaging Protocol version 4 (ICMPv4) and ICMPv6 messages. SIIT requires the assignment of an IPv4 address to the IPv6-only host, and this IPv4 address is used by the host in forming a special IPv6 address that includes this IPv4 address. The mechanism is intended to preserve IPv4 addresses; so, rather than permanently assigning IPv4 addresses to IPv6-only hosts, SIIT requires the assignment of temporary IPv4 addresses to the IPv6-only hosts. The method of assignment is beyond the scope of the SIIT RFC; the RFC suggests that Dynamic Host Control Protocol (DHCP) with short leases could be the basis for the temporary IPv4 address assignment.

The translation process can be performed directly in the end system or in a network-based device. The bump in the stack (BIS) and bump in the application (BIA), addressed in subsequent sections, are examples of translation mechanisms in end systems that employ the SIIT algorithm.

An example of a network-based translation element is shown in Figure 6.2. As indicated in the figure, upper-layer protocols such as TCP and User Datagram Protocol (UDP) can be passed through the translator relatively unscathed. For example, the SIIT translation has been designed so that UDP and TCP pseudoheader checksums are not affected by the translation process. However, the differences between

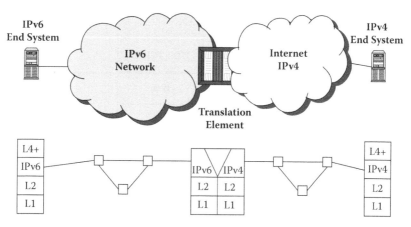

Figure 6.2   Nework-based translation example.

ICMPv4 and ICMPv6 are more significant, so SIIT specifies these translations as well (not shown in the figure). Translation can cause problems with applications like FTP that embed IP addresses in higher-layer protocols and would require the addition of application-specific application layer gateways (ALGs) in the translation element shown in the figure.

SIIT is a stateless IP/ICMP translation, which means that the translator is able to process each conversion individually without any reference to previously translated packets. Most IP header field translations are relatively simple; however, there is one issue, namely, how to translate the IP addresses between IPv4 and IPv6 packets. Translating an IPv4 address into an IPv6 address is straightforward, with SIIT defining it as embedding the IPv4 address in the low 32 bits of a specially defined IPv6 address, termed an IPv4-mapped IPv6 address. Because IPv6 addresses are much larger, some additional functionality has been defined for this mapping. In this case, SIIT uses the temporary IPv4 address assigned to the IPv6-only node and embeds this IPv4 address into a special IPv6 address, termed an IPv4-translated address. With this assignment, the IPv6-to-IPv4 address translation is also straightforward, with the result of the translation the temporary IPv4 address. The following section illustrates this process.

6.3.1.2 SIIT Details

IP Header Translation — As mentioned, the SIIT algorithm can be used as part of a solution that allows IPv6 hosts, which do not have a permanently assigned IPv4 address, to communicate with IPv4-only hosts. To handle the IP address translation between IPv4 and IPv6, SIIT defines two additional types of IPv6 addresses:

  IPv6 addresses of the type 0::FFFF:v4, termed IPv4-mapped addresses. This address is simply formed by including the IPv4 address of the IPv4 host (v4) with the prefix shown. This part of the SIIT algorithm allows for the mapping of IPv4 host addresses to IPv6 addresses.
  IPv6 addresses of the form 0::FFFF:0:v4 (note the additional 0 in the prefix), termed IPv4-translated addresses. This address is formed by including the IPv4 address temporarily assigned to the IPv6-only host and allows for the mapping of the IPv4-translated address of the IPv6 host to an IPv4 address. As the number of available globally unique IPv4 addresses becomes scarce, there is a need to take advantage of the large IPv6 address and not require that every new Internet node have a permanently assigned IPv4 address. Thus, SIIT allows for sharing the IPv4 addresses assigned to the IPv6 host.

The IP address translation process is shown in Figure 6.3. In the figure, the IPv6 host has obtained a temporary IPv4 address — $v4_{temp}$ — for use in communicating with the IPv4 host. The figure illustrates the IP address translation in going from the IPv6 host, using an IPv4 translated address, to the IPv4 host. Translation

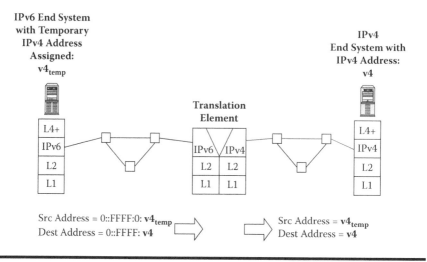

Figure 6.3   IP address translation, IPv6 to IPv4.

of the remaining fields is straightforward with a couple of exceptions (e.g., the case of an IPv6 *fragment* extension header), which are covered in detail in the RFC. If there is no IPv6 *fragment* header, the IPv4 header fields are set as follows:

- Version: 4.
- Internet Header Length: 5 (no IPv4 options).
- Type of Service and Precedence: By default, copied from the IPv6 Traffic Class (all eight bits).
- Total Length: Payload length value from IPv6 header plus the size of the IPv4 header.
- Identification: All zero.
- Flags: The More Fragments flag is set to zero, the Don't Fragment flag is set to one, and the fragment offset is all zeros.
- Time To Live (TTL): Hop limit value copied from IPv6 header, decremented by one.
- Protocol: Next header field copied from IPv6 header.
- Header Checksum: Computed once the IPv4 header has been created.

Address translation in the reverse direction is illustrated in Figure 6.4. Here again, translation of the other fields is straightforward except in the case of a packet that would require fragmentation. These packets need to be fragmented before the SIIT algorithm is applied. In the IPv4-to-IPv6 direction, the header fields are set as follows:

- Version: 6
- Traffic Class: By default, copied from IP Type of Service and Precedence field (all eight bits are copied)

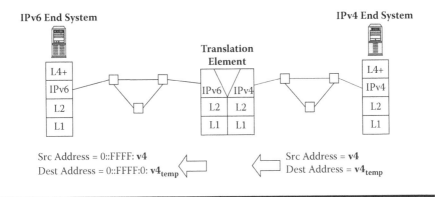

Figure 6.4   IP address translation, IPv4 to IPv6.

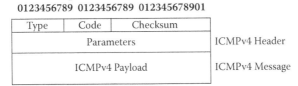

- Carried as Payload by IPv4
- Sample Query Message types
  - Echo Request and Reply (PINGS)–Types 0 and 8
- Sample Error Message Types
  - Destination unreachable–Type 3
  - Source Quench (slow down-form of flow control)–Type 4
  - Redirect or Change Route–Type 5
  - Time Exceeded (time to live expired)–Type 11
- Error Message Types Carry the IPv4 Header plus the First 64 Bits of the data field of original datagram to allow match with upper layer (eg. TCP) protocol

Figure 6.5   Internet Control Message Protocol version 4.

- Flow Label: 0 (all zero bits)
- Payload Length: Total length value from IPv4 header minus the size of the IPv4 header and IPv4 options, if present
- Next Header: Protocol field copied from IPv4 header
- Hop Limit: TTL value copied from IPv4 header decremented by one

ICMP Translation — As mentioned, ICMP is the one higher-layer protocol addressed by SIIT because there are major differences between ICMPv4 and ICMPv6.

The ICMPv4 protocol is illustrated in Figure 6.5. The Type field indicates the type of ICMPv4 message, with samples shown in the figure, and the Code field is used to provide additional information regarding the message. As examples, translating the message types shown in the figure to ICMPv6 would involve the following:

> For all translations, the Checksum field would need to be recalculated because ICMPv6, like TCP and UCP, employs a pseudoheader checksum.
>
> Echo Request and Echo Reply (Types 0 and 8) are translated to Types 128 and 129.
>
> Destination Unreachable (Type 3) is translated to Type 1. In addition to the Type field translation, translations for the Code field are also addressed in the RFC.
>
> Source quench (Type 4) has been obsoleted in ICMPv6 because it is not in current use.
>
> Redirect (Type 5) is dropped because it is valid only over a single hop.
>
> Time exceeded — time to live expired — (Type11) is translated to Type 3.
>
> As indicated in Figure 6.5, error messages contain the IPv4 header of the original IPv4 errored datagram. This IPv4 header also needs to be translated.

Translation for additional Type and Code fields is addressed in the RFC. Finally, Internet Group Management Protocol (IGMP) [RFC1112] messages are another type of ICMPv4 message. This type of message would logically be mapped to the ICMPv6 Multicast Listener Discovery (MLD) messages discussed in Section 3.2.2; however, because these are also single-hop messages, they are dropped.

Translation in the other direction, ICMPv6 to ICMPv4, generally follows the above rules but in reverse. Again, specifics of the translations are in the RFC.

6.3.2 *Bump in the Stack (BIS)*

Bump in the stack (BIS) is a variant of SIIT that performs the IPv4/IPv6 translation in the end system. With BIS, this translation occurs at the IP layer and allows an IPv4 application running on the BIS host to communicate with an IPv6 application running on a target host. BIS, described in [RFC2767], inserts modules in the end system that snoop data flowing between the IPv4 layer and the Layer 2 drivers in the host to perform the translation. Figure 6.6 illustrates the BIS architecture with the additional BIS-defined modules, *name resolver, address mapper,* and *translator*, shown in the IP layer of the end system.

During the DNS lookup of the address of the target application, the BIS modules perform a mapping of the returned IPv6 address to an IPv4 address from a pool of IPv4 addresses assigned to the modules and return this mapped IPv4 address to the application. During subsequent communication, the BIS modules perform a translation between the IPv4 and IPv6 headers using the conversion mechanism defined in SIIT.

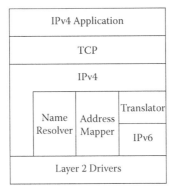

Figure 6.6 Bump-in-the-Stack architecture.

The modules operate in the following manner:

Name resolver: This module ensures that there is an IPv4 address associated with the destination application even if the application is resident on an IPv6 host. The IPv4 application on the BIS host needs such an address to allow it to communicate through its IPv4 API with the TCP layer. When the application issues a DNS query to obtain the IPv4 address of the target application, the *name resolver* intercepts this request and sends a request for both an IPv4 address and an IPv6 address. When an IPv4 address is returned, the module only needs to return this address to the application, and subsequent communication is based on IPv4. If only an IPv6 address is returned, the *name resolver* will request one of the pooled addresses from the *address mapper* module and return this address to the application. In this case, subsequent communications will be through the *translator* module.

Address mapper: The *address mapper* module maintains the list of IPv4 addresses to associate with IPv6 addresses returned to the *name resolver* from a DNS query. These IPv4 addresses are provided to the *name resolver* when requested. The module also maintains a table of the IPv6 addresses and the IPv4 addresses it has associated. This table is used to reply to requests from the translator module in performing the IPv4-to-IPv6 translation and vice versa.

Translator: This module does the actual translation between the IPv4 packets received from the BIS IPv4 application and the IPv6 packets received from the IPv6 application on the destination host.

The resulting data exchanges for the IPv4 application in the BIS host communicating with an IPv6 application are shown in Figure 6.7:

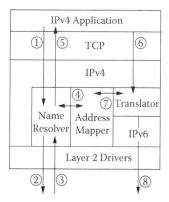

Figure 6.7 Bump in the stack information flow.

1. The IPv4 application requests the IPv4 address of the target host (e.g., www. app.com).
2. The *name resolver* intercepts the request and requests both the IPv4 and IPv6 addresses of the destination.
3. The DNS returns the IPv6 address of the destination.
4. The *name resolver* requests and receives an IPv4 address (e.g., 10.100.1.1) from the pool maintained by the *address mapper*. The *address mapper* also keeps a record of this pair of addresses.
5. The *name resolver* returns the IPv4 address (10.100.1.1) to the IPv4 application to use for subsequent communications.
6. The *translator* receives the IPv4 packet from the application.
7. The *translator* requests the IPv6 addresses associated with the IPv4 addresses in the packet and performs IPv4-to-IPv6 translation.
8. The IPv6 packet is forwarded toward the destination host.

BIS has the usual drawbacks associated with translation. In particular, IP addresses embedded by higher layers can cause problems unless special treatment is applied for each particular application. BIS does have the advantage of distributing the translation task to the end hosts and consequently may scale better.

6.3.3 Bump in the API (BIA)

6.3.3.1 Overview

Bump in the API (BIA) [RFC3338] operates in a similar way to BIS except that the translation occurs at a higher layer in the protocol suite. The main purpose of BIA is the same as BIS, namely, to allow IPv4 applications to communicate with IPv6 hosts without any modification of the IPv4 applications. However, although BIS is

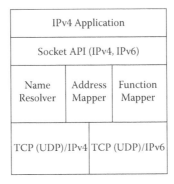

Figure 6.8   Bump-in-the-API architecture.

for systems with no IPv6 stack, BIA is for systems with an IPv6 stack, but on which some applications are not yet ported to IPv6.

The BIA technique inserts an API translator between the socket API module and the TCP/IP module in a dual-stack host so that it translates the IPv4 socket API function call into an IPv6 socket API function call and vice versa. When IPv4 applications on the dual stack communicate with other IPv6 hosts, the API translator detects the socket API function calls from IPv4 applications and invokes the IPv6 socket API function calls to communicate with the IPv6 hosts and vice versa. To support communication between IPv4 applications and the target IPv6 hosts, a pool of private IPv4 addresses is assigned via a *name resolver* function in the API translator. These addresses are assigned and mapped to the destination IPv6 address to allow the IPv4 application to conduct native IPv4 communications with the destination IPv6 host.

6.3.3.2 Details

The architecture of a BIA host is shown in Figure 6.8. The *name resolver* and *address mapper* operate in a similar manner as in the BIS architecture (refer to the BIS section for a description of these functional elements). In the BIA architecture, the *address mapper* maintains a table of pairs of an IPv4 address and an IPv6 address. The IPv4 addresses are assigned from an IPv4 address pool using unassigned IPv4 addresses (e.g., 0.0.0.1 to 0.0.0.255). The *function mapper* shown in the figure translates an IPv4 socket API function into an IPv6 socket API function and vice versa.

When detecting the IPv4 socket API functions from an IPv4 application, the *function mapper* intercepts the function call and invokes IPv6 socket API function calls that correspond to the IPv4 socket API function calls. Those IPv6 API function calls are used to communicate with the target IPv6 host. For communication from the IPv6 host to the IPv4 application, the *function mapper* works in reverse,

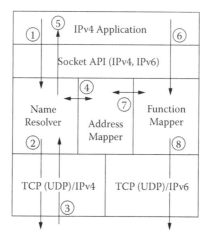

Figure 6.9 BIA call flow.

translating the IPv6 socket function call to an IPv4 socket function call to communicate with the IPv4 application.

The call flow for an IPv4 application in a BIA host communicating with an IPv6 host is shown in Figure 6.9:

1. When an IPv4 application sends a DNS query to its name server, the *name resolver* intercepts the query.
2. The *name resolver* creates a new query to resolve both A and AAAA records.
3. An AAAA record is resolved for the IPv6 host name.
4. The *name resolver* requests the *address mapper* to assign an IPv4 address from its pool corresponding to the IPv6 address.
5. The *name resolver* creates an A record for the assigned IPv4 address and returns it to the IPv4 application.
6. For the IPv4 application to send IPv4 packets, it calls the IPv4 socket API function.
7. The *function mapper* detects the socket API function from the application. In the case of IPv4 applications, the *function mapper* requests the IPv6 address from the *address mapper* associated with the IPv4 address provided in the function call. The *address mapper* returns the destination IPv6 address.
8. Using this IPv6 address, the *function mapper* invokes an IPv6 socket API function call corresponding to the IPv4 socket API function call.

Communication in the other direction, from the IPv6 host to the IPv4 application, occurs in a similar manner. When the IPv6 packet reaches the dual-stack host with the IPv4 application, the function mapper detects it. The function mapper then requests the IPv4 address associated with the IPv6 hosts address and invokes a socket IPv4 API function for communication with the IPv4 application.

6.3.4 Network Address Translation–Protocol Translation

6.3.4.1 Overview

The network address translation–protocol translation (NAT-PT) [RFC2766] employs a stateful IPv4/IPv6 header translation on a network device on the boundary of the IPv4 and IPv6 networks. The mechanism works in a similar manner as the IPv4 NAT described in [SRI199901]; the IPv4 NAT translates one IPv4 address into another. Here, the translation is between an IPv4 address and an IPv6 address. NAT-PT uses a pool of IPv4 addresses for assignment to the IPv6 nodes on a dynamic basis as sessions are initiated across the boundary and acts as a communication proxy for IPv6-only nodes communicating with IPv4 peers. One of the benefits of NAT-PT is that no changes are required to existing hosts because all the NAT-PT translations are performed at the network-based NAT-PT device.

NAT-PT differs from the SIIT algorithm in the following manner. The SIIT algorithm assumes that IPv6-only nodes are assigned an IPv4 address for purposes of communicating with IPv4 nodes. This IPv4 address is used to construct the special IPv6-translated IPv6 address, which is then used for the mapping function in the network translation device. NAT-PT uses the SIIT IP header translation algorithm for most of the header fields, but rather than assigning an IPv4 address to the IPv6-only node for use in a special address, it uses a pool of public IPv4 addresses assigned on a dynamic basis within the NAT-PT device as sessions are initiated between IPv4-only and IPv6-only nodes. A table is kept in the device of the mapping between these addresses. NAT-PT has the advantage of requiring no changes to the end systems but is stateful and does require the NAT-PT device to track ongoing sessions. All inbound and outbound IP datagrams for a session need to be routed through the NAT-PT device.

In addition to address translation, the RFC also defines network address port translation–protocol translation (NAPT-PT), which allows the multiplexing of multiple session on a single IPv4 address through use of the *port* field in upper-layer protocols such as TCP and UDP. This is similar to port multiplexing in an IPv4 environment [RFC2663].

6.3.4.2 Details

Figure 6.10, based on the RFC, illustrates the NAT-PT transition mechanism. In the figure, the following are assumed:

■ The IPv6 end system is on the same subnetwork as the NAT-PT device and uses a link-local address, FECD:BA98::7654:3210, when communicating with the NAT-PT device.
■ The session is being established by the IPv6 end system.
■ The NAT-PT device has a pool of addresses, including the subnet 120.130.26/24, to use in mapping the incoming IPv6 source addresses, in this case the above link-local address.

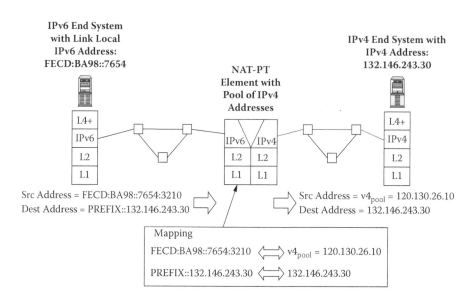

Figure 6.10   Network address translation–protocol translation data flow.

- The IPv6 domain has a prefix PREFIX::/96 assigned, and the IPv6 end system will use this prefix when addressing the IPv4 node in an IPv6 format. IPv6 packets with this prefix will be routed to the NAT-PT device. The resulting destination address is PREFIX::v4, where v4 is the IPv4 address of the IPv4 end system, 132.146.243.30 in this case.

When establishing a session from the IPv6-only node to the IPv4-only node, the IPv6 node will learn the IPv4 address of the destination IPv4 node via a DNS lookup. The information flow is then as depicted in Figure 6.10.

At session initiation, the IPv6 node will originate a packet with the following:

IPv6 source address: FECD:BA98::7654:3210

IPv6 destination address: PREFIX::132.146.243.30.

On reception of the packet, the NAT-PT device will assign an IPv4 address from its pool, and this assigned address will be used as the source address in forwarding the packet to the IPv4 node. The resulting translated packet will have the following:

IPv4 source address: 120.130.26.10, assigned for the IPv4 address pool
IPv4 destination address: 132.146.243.30, the IPv4 address of the IPv4 end system

This IPv6-to-IPv4 mapping is retained for the duration of the session. Based on this retained mapping, returning traffic will be recognized by the NAT-PT device as belonging to the same session. Because NAT-PT maintains translation state, each session must be routed through the same NAT-PT device.

In the above case, the session was initiated by the IPv6 end system, and the allocation of the IPv4 address from the pool was triggered by the first IPv6 packet of the session. When the session is initiated by the IPv4 end system, the allocation process is different. In this case, the IPv4 end system will perform a DNS lookup for the IP address of the IPv6 end system, and the lookup will be routed to a DNS application layer gateway (DNS-ALG) colocated in the NAT-PT device. The DNS-ALG will change the query to request an IPv6 address record. When the IPv6 address record is returned, the DNS-ALG will intercept the response, obtain an IPv4 address from the NAT-PT pool, and reply to the IPv4 end system with the IPv4 address record for the pooled IPv4 address.

6.3.5 Transport Relay Translator

6.3.5.1 Overview

Transport layer relays, such as those found in firewall/proxy products, can also be functionally extended into IPv6/IPv4 translators. This is a natural extension of these products because they already act at the transport layer and higher layers in providing security services for corporate networks. For a transport layer relay, a relay process on a network border device (e.g., a router) partitions the transport layer path into two "terminated" IP legs for which one leg utilizes IPv6 and one leg IPv4. TCP/UDP segments traversing the relay pass up to the transport layer, are processed, and then get sent out in the adjacent leg. Translation only occurs at Layer 4; therefore, IP layer conversion is avoided. The IPv6-to-IPv4 header converters discussed previously, such as BIS, have to take care of IP layer issues involving path MTU and fragmentation; however, because transport relay translator (TRT) operates above the IP layer, it is free from such problems.

In the case of TCP, there are two TCP layer connections established, one TCP/IPv6 and the other TCP/IPv4. Some TCP segment processing thus needs to be performed in the relay. For example, because there are two independent TCP connections involved, the sequence number and acknowledgement fields of the segments need to be mapped between the connections. This is thus a stateful translation, and traffic between peers needs to pass through the same transport relay element.

The TRT [RFC3142] assumes that traffic is initiated by the IPv6-only host destined to an IPv4-only host. The mechanism is thus useful for initial deployment of an IPv6-only network that needs access to legacy IPv4-only network resources. TRT is designed to require no modification on the IPv6-only initiating hosts or on the IPv4-only destination hosts.

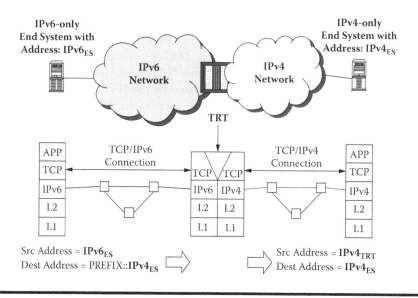

Figure 6.11 Transport relay translator (TRT) flow for TCP.

Communication with the TRT element is initiated from the IPv6 side via a special destination address type (a 64-bit prefix along with the IPv4 address of the destination node inserted in the last 32 bits of the address). Routing information in the IPv6 network is configured to route this prefix toward the dual-stacked TRT router which terminates the IPv6 session and initiates IPv4 communication to the final destination. The mechanism can be extended to handle sessions initiated in the opposite direction, but this extension is not addressed in the RFC.

6.3.5.2 Details

TRT communications for TCP is depicted in Figure 6.11. When the initiating host (whose IPv6 address is $IPv6_{ES}$) wishes to make a connection to the destination host, it first resolves the destination host's IP address, in this case $IPv4_{ES}$. As mentioned, the initiating host must use a special form of IPv6 address to connect to the IPv4 destination host. The special form of address, as shown in the figure, is PREFIX:: $IPv4_{ES}$, for which PREFIX::/64 is a part of the IPv6 unicast address space assigned to the IPv6 network. Routing information in the IPv6 network is configured so that packets destined to PREFIX::/64 are routed toward the TRT system. Thus, when the initiating IPv6 host sends a TCP connection establishment segment and future TCP session segments to the destination host, the TCP packets are routed toward the TCP relay system based on the routing configurations. The TCP relay system receives and accepts the packets even though the TCP relay system does not own the destination IPv4 address. The IPv6 layer is terminated in the TRT device, and the TRT relays TCP traffic on a specific port between the TCP layers.

In the figure, the IPv6-only host has established a TCP connection with the TRT network element, and the TRT element has established an independent TCP connection with the IPv4-only host. IPv6 packets from the IPv6-only host will have the source address of the IPv6-only host ($IPv6_{ES}$) and a destination address of the special form $PREFIX::IPv4_{ES}$. Based on the routing tables for the PREFIX::/64 packets, the packet is routed to the TRT. The TRT system investigates the low-ermost 32 bits of the destination address in the datagram and incorporates this IPv4 address as the destination IPv4 address of the relayed segment. The source address of the relayed IPv4 packet is the IPv4 address of the TRT relay device interface, $IPv4_{TRT}$. IP packet flow in the reverse direction is similar, with the TRT relay device providing a mapping of the source and destination IP addresses on the incoming packet based on the incoming TCP port number: $IPv4_{ES}$ to $PREFIX::IPv4_{ES}$ and $IPv4_{TRT}$ to $IPv6_{ES}$.

Like any translator that modifies the IP address, a TRT system may need to modify data content for application protocols that contain an embedded IP address.

For sessions initiated by the IPv4-only host, translation is more difficult to support. This situation is not defined in the RFC and would require nontrivial mapping between DNS names to temporary IPv4 addresses, similar to what is discussed in the NAT-PT section.

6.4 Tunneling

Tunneling is an often-used technique by which one transport protocol is encapsulated as the payload of another. This allows for the transport of the encapsulated data unit across the encapsulating protocol's transport network. Typically, when employed as part of an IPv6 transition mechanism, the existing IPv4 transport infrastructure is used to encapsulate IPv6 packets, thereby using the existing IPv4 infrastructure to provide basic IPv6 connectivity. This encapsulation of IPv6 packets in IPv4 packets (using IP Protocol Type 41) is illustrated in Figure 6.12.

Figure 6.13 is an example of the protocol architecture for IPv6 encapsulation in IPv4. The technique works as follows: The egress points of the linked networks encapsulate IPv6 packets to specified IPv6 destinations through IPv4 interfaces. The packets proceed over the normal IPv4 routing system and are decapsulated at

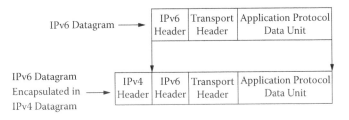

Figure 6.12   IPv6 encapsulation in IPv4.

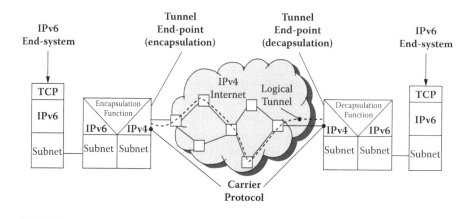

Figure 6.13 Protocol architecture for IPv6 tunneling.

the other end of the tunnel, with the IPv6 packet then forwarded to the correct host by the IPv6 routing system. A number of methods have been defined to determine the endpoint of the encapsulating tunnels, including static configuration in the encapsulating/decapsulating systems and automatic tunneling techniques, generally dependent on special forms of IPv6 addressing.

6.4.1 Static Tunneling

Static tunneling can be used to link isolated islands of IPv6, in which the network domains are well known and unlikely to change without notice. This technique is most useful when just a few tunnels are needed (e.g., interconnecting a few large IPv6 networks).

A static configured tunnel is equivalent to a permanent link between two IPv6 domains with the permanent connectivity provided over an IPv4 backbone. For static tunneling, assigned IPv4 addresses are manually configured to the tunnel source and the tunnel destination. The determination of which packets to tunnel is made via a routing table in the tunnel endpoints; the table directs packets based on their destination address using the prefix mask-and-match technique. Note that the host or router at each end of a static configured tunnel must support both the IPv4 and IPv6 protocol stacks.

6.4.2 Automatic Tunneling Using IPv4-Compatible Addresses

[RFC2893] describes the notion of an automatic IPv4 tunnel. In this technique, the prefix ::/96 is set aside for IPv4-compatible addresses, for which the rightmost 32 bits of the IPv6 address are the IPv4 address of a recipient node. IPv6 packets addressed to these IPv6 addresses are automatically encapsulated in an IPv4 packet

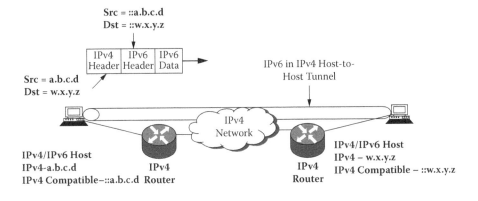

Figure 6.14 Automatic tunneling with IPv4-compatible IPv6 addresses.

addressed to the corresponding IPv4 address and tunneled to its destination. Figure 6.14 illustrates this encapsulation process.

The host or router at each end of an IPv4-compatible tunnel must support both the IPv4 and IPv6 protocol stacks. Routing tables are set such that ::/96 addresses are directed to the automatic tunnel interface.

The IPv4-compatible address scheme is fairly restricted. Because no further routing occurs beyond the tunnel endpoint, the tunnel endpoint is also the recipient. This technique does not scale to large networks and is deprecated in favor of other solutions, such as 6to4 and Intrasite Automatic Tunnel Addressing Protocol (ISATAP), addressed in other sections.

6.4.3 6over4 Transition Mechanism

The 6over4 transition mechanism (also known as IPv4 multicast tunneling) [RFC2529] uses the multicast capability of an IPv4 network to allow the IPv4 network to act as a virtual subnet for IPv6 hosts and routers. The mechanism essentially uses IPv4 as the Layer 2 transport for IPv6. This mechanism behaves much like an IPv6-over-Ethernet subnetwork except that the Layer 2 functionality of Ethernet is replaced by that of the multicast IPv4 network. This mechanism differs from the 6to4 mechanism discussed in the next section in that it allows full neighbor discovery with the IPv4 network acting as a virtual local area network (LAN). Because 6over4 defines the IPv4 multicast network as just another Layer 2 media type for IPv6, this mechanism does not require any special prefix as is the case for the 6to4 mechanism (Section 6.4.4).

Because 6over4 treats the IPv4 infrastructure as a single link with multicast capabilities, the Neighbor Discovery processes (such as address resolution and router discovery) work as they do over any physical link with multicast capabilities. This relationship is shown pictorially in Figure 6.15 and Figure 6.16, which depict the physical and logical architectures for the 6over4 transition scheme, respectively.

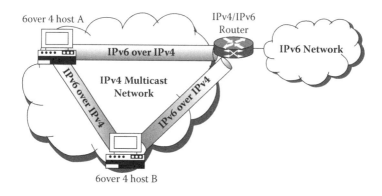

Figure 6.15   Physical architecture for 6over4 network.

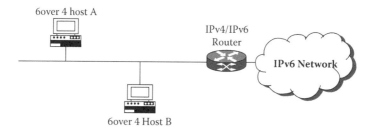

Figure 6.16   Logical view of 6over4 network.

For efficient operation, a 6over4 host needs to form a number of addresses. In forming these addresses, a 6over4 node uses the IPv4 address of the interface in the same way that a node with an Ethernet interface uses the 64-bit extended unique identifier (EUI-64) interface ID:

- Unicast address: A unicast address is formed in the following manner: 6over4 hosts use a valid 64-bit prefix for unicast addresses and the 64-bit interface identifier ::$IPv4_{add}$, where $IPv4_{add}$ is the 32-bit IPv4 address assigned to the host.
- Link-local address: By default, 6over4 hosts automatically configure the *link-local* address FE80::$IPv4_{add}$ on each 6over4 interface. For example, a node with address 10.0.0.1 will end up with link-local address FE80::0A00:0001.
- Solicited-node multicast address: For efficient address resolution, the node is assigned a solicited-node multicast address for each of its unicast addresses. IPv6 uses this solicited-node multicast address when sending a neighbor solicitation message as part of address resolution, and it has the advantage of not disturbing hosts with address resolution messages (which occur often on a link) that are not intended for them. The solicited-node multicast address

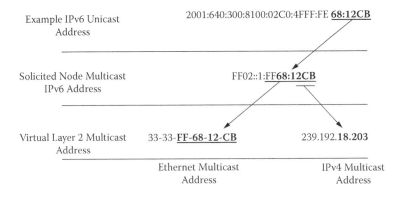

Figure 6.17   Ethernet and IPv4 multicasting scheme for IPv6.

is used in the following manner: Instead of sending the neighbor solicitation message to every node on the local link (via an FF02::1 all-nodes multicast address), the neighbor solicitation message is sent to a very restricted multicast group identified by the nodes solicited-node multicast address. The solicited-node multicast address for a given unicast address is constructed by taking the last three octets of the unicast address and prepending FF02::1:FF00:0000/104. Thus, the solicited-node multicast address of unicast address 2001:630:200:8100:02C0:4FFF:FE68:12CB is FF02::1:FF68:12CB, as illustrated in the upper part of Figure 6.17. It is this solicited-node multicast address that a node uses as the destination of a neighbor solicitation packet.

■ To incorporate the multicasting capability of the IPv4 network, a final mapping is required between the solicited-node multicast address and the IPv4 multicast address used when sending the neighbor solicitation message. Here again, a process analogous to that defined for IPv6 multicast over Ethernet is employed. The mappings for both Ethernet and IPv4 are shown in the lower portion of Figure 6.17. For Ethernet, the multicast destination address is formed by prepending 33-33- (indicating multicast) to the last 32 bits of the IPv6 solicited-node multicast address. In the case of 6over4, an IPv6 solicited-node multicast destination address is mapped to the IPv4 multicast address using the IPv4 multicast address taken from the block 239.192.0.0/16, a subblock of the organization-local scope address [RFC3171] and appended to this are the last 16 bits of the solicited-node multicast address.

With the above-defined mappings, the IPv6 Neighbor Discovery procedures defined in [RFC2461] can be followed. As discussed in Section 3.2.2., these procedures involve the exchange of router solicitation, router advertisement, neighbor solicitation, neighbor advertisement, and redirect ICMP messages.

An example of this process is shown in Figure 6.18, which illustrates a neighbor solicitation message requesting the IPv4 link-layer address of the target node while

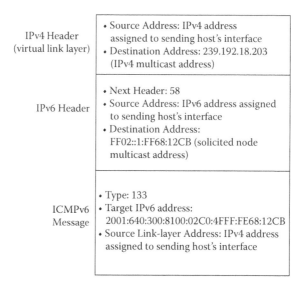

IPv4 Header (virtual link layer)	• Source Address: IPv4 address assigned to sending host's interface • Destination Address: 239.192.18.203 (IPv4 multicast address)
IPv6 Header	• Next Header: 58 • Source Address: IPv6 address assigned to sending host's interface • Destination Address: FF02::1:FF68:12CB (solicited node multicast address)
ICMPv6 Message	• Type: 133 • Target IPv6 address: 2001:640:300:8100:02C0:4FFF:FE68:12CB • Source Link-layer Address: IPv4 address assigned to sending host's interface

Figure 6.18 Illustrative neighbor solicitation message.

also providing the node's own IPv4 link-layer address to the target. Figure 6.18 is an example of a neighbor solicitation message sent when a host is resolving the IPv6 unicast address shown in Figure 6.18. Only the key header fields are shown in the figure.

Note the following:

- The destination IPv4 address (239.192.18.203) is the associated IPv4 multicast address derived from the unicast address being resolved. The target host will receive this IPv4 packet because it is a member of this multicast group. The source IPv4 address is that of the interface of the host requesting the address resolution.
- IPv6 fields include a source address that is the address assigned to the interface from which this message is sent, a destination address that is the solicited-node multicast address corresponding to the target address, as shown in Figure 6.17, and a Next Header field with value 58 indicating ICMPv6.
- ICMPv6 fields include a Type field with value 135 indicating a neighbor solicitation message, the IPv6 address of the target of the solicitation, and the source link-layer address, in this case, the IPv4 address assigned to the host requesting the address resolution.
- Neighbor solicitations are multicast when the node needs to resolve an address and unicast when the node seeks to verify the reachability of a neighbor.

It should be noted that, because 6over4 relies on IPv4 multicast, a feature not widely deployed, the 6over4 mechanism is not expected to be widely used.

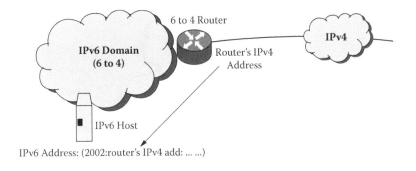

Figure 6.19   6to4 automatic tunneling.

6.4.4 *6to4 Transition Mechanism*

6.4.4.1 *Overview*

The 6to4 transition mechanism provides for the interconnection of isolated IPv6 domains via automatic tunnels through an IPv4 Internet. The motivation for this method is to allow isolated IPv6 domains or single hosts attached to an IPv4 network to communicate with other such IPv6 domains or hosts with minimal manual configuration. The mechanism also allows connection of these domains to the IPv6 Internet over the IPv4 Internet. Effectively, it treats the IPv4 Internet as a unicast point-to-point link layer for the purposes of interconnecting the isolated IPv6 domains with one another and with the IPv6 Internet.

Automatic tunneling in this case is accomplished by having a router, termed a 6to4 router, on the border of the isolated IPv6 domain and attached to the IPv4 Internet. The 6to4 mechanism operates by having the IPv4 address of this router's IPv4 interface be a portion of the prefix of the IPv6 addresses assigned to the IPv6 host in the respective IPv6 domain. Thus, specifying an IPv6 host's 6to4 IPv6 address explicitly identifies the IPv4 tunnel endpoint of its border 6to4 router. This process is illustrated in Figure 6.19, in which the IPv6 address of the host in the 6to4 domain contains the IPv4 address of the border router. Additional details of the 6to4 addressing scheme are discussed in the next section.

Figure 6.20 illustrates the 6to4 transition mechanism for interconnecting IPv6 6to4 domains. The figure depicts two isolated 6to4 networks. Each site has a router configured with a connection to the IPv4 network and the ability to create automatic 6to4 tunnels across the IPv4 network to interconnect the 6to4 sites.

Note the following with respect to the 6to4 transition mechanism:

1. When the 6to4 border router of a 6to4 domain is routing an IPv6 packet to a host in another 6to4 domain shown in the figure, the router can automatically determine the tunnel endpoint from the IPv4 address embedded in the IPv6 6to4 address.

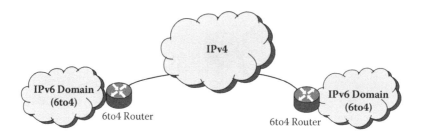

Figure 6.20 Interconnecting 6to4 domains.

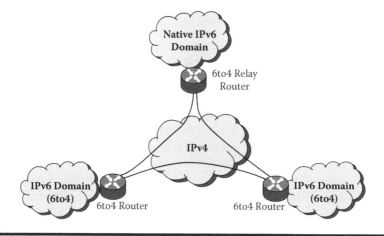

Figure 6.21   Relay router: Interconnecting 6to4 domain with native IPv6 domain.

2. This mechanism conserves IPv4 addresses because only one globally unique unicast IPv4 address is needed for an entire IPv6 domain.
3. This mechanism falls into the category of a router-to-router tunneling transition mechanism.

In addition to interconnecting 6to4 domains, the 6to4 mechanism also allows for the interconnection of these domains to native IPv6 networks via what are termed relay routers. Relay routers are essentially bridges between the 6to4 sites and native IPv6 domains. Figure 6.21 illustrates both the interconnection of the 6to4 domains and the interconnection of these domains with the native IPv6 network.

The major elements of the 6to4 transition mechanism are the following:

■ 6to4 host: Any IPv6 host that is configured with at least one 6to4 IPv6 address.
■ 6to4 router: An IPv6/IPv4 router that supports the use of a 6to4 tunnel interface and is typically used to forward 6to4-addressed traffic between the 6to4 hosts within a site and other 6to4 routers or 6to4 relay routers over an IPv4

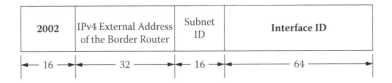

Figure 6.22   6to4 addressing scheme.

Internet. 6to4 routers require additional processing logic for proper encapsulation and decapsulation and might require additional manual configuration.

- 6to4 relay router: An IPv6/IPv4 router that forwards 6to4-addressed traffic between 6to4 routers on the Internet and hosts on the IPv6 Internet.

6.4.4.2 6to4 Addressing and Site Routing

[RFC3056] defines a method for assigning a unique IPv6 address prefix for the hosts in a 6to4 domain. For this purpose, the IANA has permanently assigned a specific 6to4-reserved IPv6 address space: 2002::/16. The remainder of the 6to4 prefix is obtained by appending the 32-bit IPv4 address assigned to the 6to4 router external IPv4 interface shown in Figure 6.19. The resulting 6to4 address is illustrated in Figure 6.22. The 6to4 prefix is often abbreviated as 2002:V4ADDR::/48.

Within the IPv6 domain, the prefix can be used exactly like any other valid IPv6 prefix, such as for automated address assignment and discovery according to the normal mechanisms discussed in Section 3.2.2. Local IPv6 routers within the site advertise 2002:V4ADDR:SubnetID::/64 prefixes so that hosts can create an autoconfigured 6to4 address with this prefix. Within the local routers, 64-bit prefix routes are used to route traffic within the site, and traffic that does not match the 64-bit prefix of one of the site's subnets is forwarded to the site's 6to4 border router. When an outgoing packet reaches the 6to4 router, it is encapsulated as discussed in the next section.

6.4.4.3 6to4 Transition Mechanism Details

Figure 6.23 illustrates the flow of packets from a host in one 6to4 domain to a host in another 6to4 domain. The 6to4 domain on the left has the 6to4 prefix of 2002:C251:2E01::/48, derived from the 2002::/16 IPv6 prefix and the IPv4 address of the border router: 194.81.46.1. Derivation of this 6to4 prefix is shown in the figure. Similarly, the 6to4 domain on the right has the 6to4 prefix of 2002:C253:6A06::/48, with the derivation also shown in the figure. The IPv6 host on the left has the 6to4 IPv6 address of 2002:C251:2E01:1::1, indicating Host 1 on Subnet 1 of the 6to4 domain. The IPv6 host on the right has the 6to4 IPv6 address of 2002:C253:6A06:2::2, indicating Host 2 on Subnet 2 of the 6to4 domain.

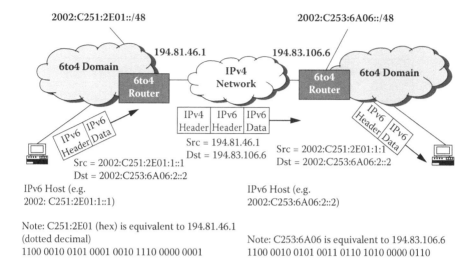

Figure 6.23 Communication between 6to4 domains.

When the host on the left resolves the name of the host on the right, it forwards an IPv6 packet as shown toward the 6to4 router. The 6to4 router has a 2002::/16 route that is used to forward traffic to other 6to4 sites. This route forwards traffic to a router interface, termed a pseudointerface in the RFC, that encapsulates the IPv6 packet in an IPv4 packet (with protocol field 41 indicating encapsulation). The source address is that of the 6to4 border router, and the destination address is derived from the IPv6 address of the destination host (which has the address embedded as part of its IPv6 6to4 address). The packet is then forwarded to the 6to4 router at the destination 6to4 domain, where it is decapsulated, and using the 64-bit prefix in its routing table, the IPv6 payload is forwarded toward the destination host.

In addition to communicating with hosts in other 6to4 domains, a host in a 6to4 domain will also need to communicate with hosts in a native (non-6to4) IPv6 network. To allow hosts using 6to4 addresses to exchange traffic with hosts using native IPv6 addresses, relay routers, shown in Figure 6.21, have been defined. These relay routers connect to both the IPv4 and the IPv6 networks. A relay router will advertise 2002::/16 into the IPv6 network so that packets with this prefix will arrive on the IPv6 interface and be forwarded over the IPv4 network via a 6to4 tunnel. Packets arriving at the relay router on an IPv4 interface will have their IPv6 payloads forwarded to the IPv6 network.

Referring to Figure 6.21, for a 6to4 router to allow communication with hosts on the native IPv6 Internet, it must have its default route, ::/0, set to a 6to4 address that contains the IPv4 address of a 6to4 relay router. To avoid the need for users to set this up manually, [RFC3068] has defined an IPv4 6to4 relay anycast address of 192.88.99.1, with a 6to4 anycast prefix of 192.88.99.0/24, for the purpose of sending packets to a relay router. The 6to4 relay routers advertise this 6to4 anycast prefix

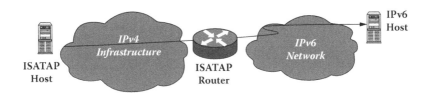

Figure 6.24   ISATAP architecture.

using the routing protocol of the IPv4 network. According to the 6to4 addressing rules, the default route will then point to 2002:c058:6301::, containing the embedded anycast address.

6.4.5 *Intrasite Automatic Tunnel Addressing Protocol (ISATAP)*

6.4.5.1 *Overview*

The primary function of ISATAP [RFC4214] is to allow hosts that are multiple IPv4 hops away from an IPv6 router to participate in the IPv6 network by automatically tunneling IPv6 packets over IPv4 to the IPv6 router as the next-hop address. ISATAP connects the isolated IPv6 nodes within IPv4 sites via automatic IPv6-in-IPv4 tunnels. Although the ISATAP tunneling mechanism is similar to other automatic tunneling mechanisms, such as 6over4 tunneling, ISATAP is designed for transporting IPv6 packets within a site where a native IPv6 infrastructure is not yet available. ISATAP is similar to the 6over4 mechanism in that it uses IPv4 as a link layer to interconnect IPv6 nodes and tunnels the IPv6 datagram over IPv4, creating a virtual link over the IPv4 network. The major difference is that, unlike 6over4, ISATAP does not assume a multicast-capable IPv4 infrastructure. It assumes the IPv4 infrastructure is a nonbroadcast multiple access (NBMA) subnetwork, and the characteristics of such a network are taken into account when specifying functions such as neighbor and router solicitation. The ISATAP architecture is illustrated in Figure 6.24.

As with many of the IPv6 transition schemes, ISATAP addressing is key to its proper functioning. For example, to avoid the need for Neighbor or Router Discovery, the interface ID portion of an ISATAP address contains the IPv4 address of the host or router interface. Thus, when communicating directly with a neighbor, the neighbor's link-local address will automatically indicate the tunnel endpoint in the interface ID portion of the address. Besides the IPv4 address, the interface ID contains a special value for the remaining 32 bits to indicate an ISATAP host or router.

6.4.5.2 *ISATAP Addressing*

Figure 6.25 shows the structure of an ISATAP address. Like other transition schemes, such as IPv4-compatible addressing, 6over4, and 6to4, ISATAP addresses contain an embedded IPv4 address. With ISATAP, the IPv4 address is that assigned

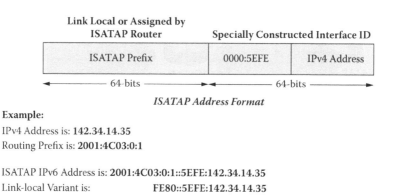

ISATAP Address Format

Example:

IPv4 Address is: **142.34.14.35**
Routing Prefix is: **2001:4C03:0:1**

ISATAP IPv6 Address is: **2001:4C03:0:1::5EFE:142.34.14.35**
Link-local Variant is: **FE80::5EFE:142.34.14.35**

Figure 6.25   ISATAP address structure.

to the ISATAP host or router interface. This embedded address can thus be used as the tunnel endpoint when ISATAP traffic is tunneled over an IPv4 infrastructure. In addition to the IPv4 address, the interface identifier has the special ISATAP 32-bit identifier sequence 0000:5EFE. ISATAP thus uses addresses with an interface ID of ::0:5EFE:w.x.y.z, which is assumed to correspond to an IPv4 "link-layer" address of w.x.y.z. ISATAP host and routers are also automatically configured with the link-local address of FE80::5EFE:w.x.y.z.

RFC 4214 specifies a number of methods to account for the lack of host's access to multicast for router solicitation. Methods suggested include manual configuration of a router's IPv4 address in a host table, a DHCPv4 option, or a DNS lookup for a fully qualified domain name (e.g., by looking up a host name such as isatap. router.com). Once the host has the IPv4 addresses of potential ISATAP routers, it can send router solicitations to each, using the router's link local address and tunneling the solicitation in IPv4. The routers can reply, and the nodes can form an RFC 2373-compliant IPv6 globally aggregatable unicast address based on the advertised prefixes and their derived ISATAP interface IDs [RFC2373].

Figure 6.25 shows what ISATAP addresses would look like if the advertised prefix were 2001:4C03:0:1::/64 and the IPv4 address was 142.34.14.35. In the ISATAP address, the IPv4 address is expressed in dotted-decimal notation. ISATAP hosts and routers thus generally have three types of addresses: IPv6 globally aggregatable unicast, IPv6 link-local, and IPv4 addresses.

6.4.5.3 Example Network

For communications between ISATAP nodes, a node will know that the destination is an ISATAP node based on the interface identifier. Based on the prefix, if the destination is onlink, the IPv6 packet is encapsulated in an IPv4 packet, and the destination IPv4 address is derived from the IPv4 address embedded in the link-local destination IPv6 ISATAP address. If the destination is offlink, then the

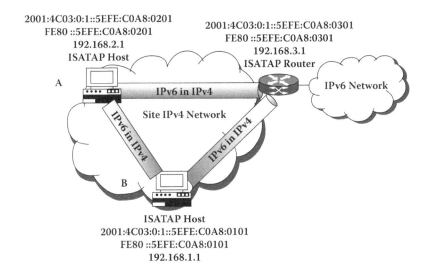

Figure 6.26   Sample ISATAP network.

IPv6 packet is still encapsulated in IPv4, and the destination is the current default ISATAP router that would forward the IPv6 packet toward the destination, either via a direct connection to the IPv6 network or a further IPv4 tunneling mechanism such as 6to4.

An example of an ISATAP network is shown in Figure 6.26. Hosts A and B and the ISATAP router have private IPv4 site addresses assigned to their interfaces along with IPv6 global-scope and link-local ISATAP addresses. When Hosts A and B exchange IPv6 traffic using their link-local addresses, the traffic is sent via an IPv6-in-IPv4 tunnel with endpoints that are derived from the interface ID portion of the link-local addresses.

For offlink traffic, which is traffic not corresponding to the prefix 2001:4C03:0:1, Hosts A and B have a default route, pointing to the ISATAP address that corresponds to the interface of the ISATAP router. In the figure, the ISATAP router is configured with the IPv4 address of 192.168.3.1, and thus Hosts A and B are configured with a default route (::/0), which uses the ISATAP address of FE80::5EFE:C0A8:0301 as the next-hop address, where C0A8:0301 is the hexadecimal value of the IPv4 address. As a result, all IPv6 traffic that matches this default route is encapsulated and forwarded to the ISATAP router, which then forwards the traffic.

6.4.6 Teredo

6.4.6.1 Overview

Teredo is an IPv6 transition technology that provides address assignment and host-to-host automatic tunneling for unicast IPv6 traffic when IPv6/IPv4 hosts

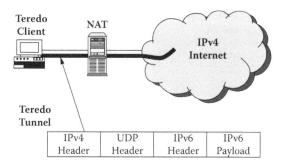

| IPv4 | UDP | IPv6 | IPv6 |
| Header | Header | Header | Payload |

Figure 6.27   Teredo client and tunnel.

are located behind one or multiple IPv4 network address translators (NATs). The basic NAT operation was defined in [RFC1631] with the intent of conserving IPv4 addresses and involves a mapping between private, internal IPv4 addresses and port numbers within a subnetwork to public, external IP addresses and port numbers assigned by the NAT device. To traverse IPv4 NATs, Teredo specifies IPv6 packets sent as IPv4-based UDP messages. Teredo also builds on the techniques defined in [RFC3489] for tunneling UDP traffic through various types of NATs.

A basic diagram for a host implementing the Teredo client is shown in Figure 6.27. Here, an NAT is used to provide Internet connectivity for this site, an implementation that is typical for a small office/home office (SOHO) environment. Other elements of the Teredo architecture are addressed in the following sections. The Teredo packet is also shown in the figure.

Like the 6to4 mechanism, Teredo is an automatic tunneling technology but differs from 6to4 in a number of aspects. For example, unlike the 6to4 mechanism, in which the automatic tunnel originates in the 6to4 edge router and IPv6 is the subnet technology, the Teredo tunnel originates at the host and uses IPv4 as the subnet technology to route to the NAT device.

NAT devices also cause problems for the 6to4 mechanism. 6to4 relies on the 6to4 routing functionality being implemented in the network connectivity device, a functionality that is not common for SOHO NAT devices. When the device did implement the 6to4 functionality, the 6to4 function requires the assignment of a public IP address, which is not possible in cases involving multiple levels of NATs. In addition, NAT devices can usually only deal with TCP, UDP, and limited ICMP messages. 6to4 tunnels make use of IPv4 Protocol Type 41, which means that it may not be possible for NATed networks to use 6to4 or indeed any other mechanisms using protocol types differing from those of TCP, UDP, or ICMP.

Much of the Teredo specification is concerned with identifying the specific type of NAT deployed in a network and specifying procedures for handling these various types. Examples of the types of NATs include the following, analyzed in [RFC3489] (these various types of NATs are depicted pictorially in Figure 6.28):

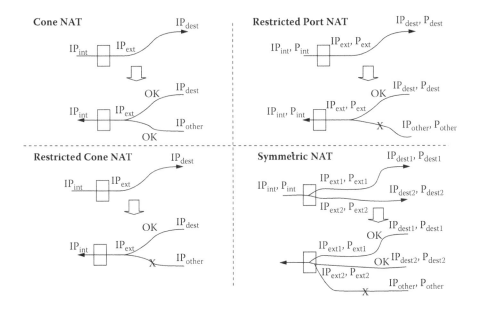

Figure 6.28 Types of NATs.

Cone NAT: The NAT device maintains a one-to-one mapping between an internal address and port number and an external address and port number. In addition, any external host can send a packet to the internal host by sending a packet to the mapped external address.

Restricted cone NAT: Similar to a cone NAT except that an external host can send a packet to the internal host only if the internal host had previously sent a packet to the external host (essentially, the previous packet opens up the NAT for the external host).

Port restricted cone NAT: A port restricted cone NAT is similar to a restricted cone NAT, but the restriction includes port numbers. An external host can send a packet, with source IP address X and source port P, to the internal host only if the internal host had previously sent a packet to IP address X and port P (essentially, the previous packet opens up the NAT for a specific port on the external host).

Symmetric NAT: Instead of a one-to-one mapping between internal IP address and port number to external address and port number, a different external address and port number are assigned for each external host. In addition, only external hosts that received a previous packet from the internal host may send to the internal host.

[RFC4380] also defines a bubble packet, used to create a mapping in NAT devices. The bubble is a minimal IPv6 packet consisting of an IPv6 header with

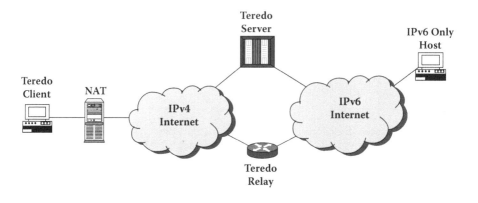

Figure 6.29   Teredo architecture.

no payload and can be used to open up the NAT for communication through the various types of NATs shown in Figure 6.28.

6.4.6.2 Architecture

The Teredo architecture is shown in Figure 6.29. Major components in the architecture are the following:

■ Teredo client: A node that has access to the IPv4 Internet and wants access to the IPv6 Internet. The node supports a Teredo UDP tunneling interface through which packets are tunneled to either other Teredo clients or nodes on the IPv6 Internet (via a Teredo relay). A Teredo client communicates with a Teredo server to obtain an address prefix from which a Teredo-based IPv6 address is configured. The IPv6 Teredo address structure and configuration procedure are discussed in a following section.

■ Teredo server: A node that is used to assist in the provision of IPv6 connectivity to Teredo clients. This node supports a Teredo tunneling interface over which packets are received from Teredo clients. The general role of the Teredo server is to assist in the address configuration of Teredo clients and to facilitate the initial communication between Teredo clients and other Teredo clients or between Teredo clients and IPv6-only hosts. The Teredo server listens on UDP port 3544 for Teredo traffic, for example, responding to router solicitation messages from Teredo clients with router advertisements to assist in Teredo address configuration (discussed in Section 6.4.6.3).

■ Teredo relay: An IPv6 router that can receive traffic from the IPv6 Internet destined for the Teredo client and forward it to the Teredo client interface. Teredo relays also accept packets sent by Teredo clients over their Teredo interface for forwarding to the IPv6 Internet.

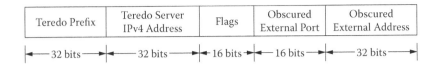

Figure 6.30 Teredo address format.

6.4.6.3 *Teredo Addressing and Address Configuration Process*

The Teredo address format is shown in Figure 6.30:

- The Teredo prefix is assigned the value 2001::/32 in [RFC4380].
- The Teredo server IPv4 address is the public IPv4 address of the server that assisted in configuring the Teredo IPv6 address for the client.
- The flag field is derived during the Teredo address configuration procedure and indicates the type of NAT used by the Teredo client.
- The last two fields are the "obscured" mapped external IPv4 address and port of the Teredo client. RFC 4380 defines obscuring as reversing each bit in the address and port fields. This obscuring is necessary because many NAT devices perform a deep packet inspection process and could perform an undesired mapping of the external addresses and ports.

Teredo address configuration is achieved via interactions between the Teredo client and server. The client and server exchange router solicitation and router advertisement messages, allowing the client to determine the 64-bit prefix (from the received router advertisement), the type of NAT involved, and the external IP address and port number (also learned from the router advertisement).

6.4.6.4 *Sample Teredo Communication*

RFC 4380 addresses the communication procedures for a Teredo client when communicating in a number of different situations. It addresses cases of differing types of NATs, communications with other Teredo clients, and communications with IPv6-only hosts on the IPv6 Internet. As an illustrative example, we show the case of a Teredo client communicating with a IPv6-only host for the case of a cone NAT in Figure 6.31:

1. To determine the Teredo relay serving the IPv6-only host, the client sends an IPv6 echo request via its Teredo server. Teredo servers are expected to relay these requests.
2. The Teredo server relays the echo request to the IPv6-only host.
3. The IPv6-only host sends an IPv6 echo reply with the Teredo client address as destination. The IPv6 infrastructure will route this packet to the nearest Teredo relay based on 2001::/32 routes.

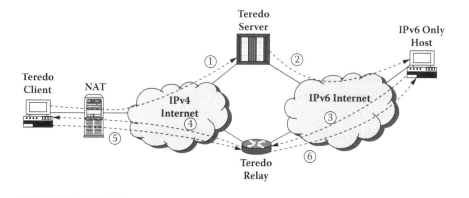

Figure 6.31   Sample Teredo communication with IPv6-only host and cone NAT.

4. The Teredo relay will tunnel the echo reply to the Teredo client, extracting the external IP address and port number from the Teredo address. In the case of a cone NAT, the packet will be forwarded to the Teredo client. Note that in the case of a restricted cone NAT, this packet would be discarded, and additional procedures, involving bubble packets, would be required to ensure the NAT is opened for this communication.
5. The client determines the relay IPv4 address from the received packet and can begin sending packets to the IPv6-only host via the Teredo relay.
6. The relay extracts the IPv6 packet and forwards to the IP-only host. Future communications can follow this same path.

Additional examples of Teredo communications can be found in [MIC200601].

References

[MIC200601] Microsoft Corporation, Toredo Overview, October 30, 2006, http://www.microsoft.com/technet/prodtechnol/winxppro/maintain/teredo.mspx.

[MIC200701] Changes to IPv6 in Windows Vista® and Windows Server® "Longhorn," January 2, 2007, http://www.microsoft.com/technet/community/columns/cableguy/cg1005.mspx.

[RFC1112] S. Deering, Host Extensions for IP Multicasting, RFC 1112, August 1989.

[RFC1631] K. Egevang, P. Francis, The IP Network Address Translator (NAT), RFC 1631, May 1994.

[RFC2373] R. Hinden and S. Deering, IP Version 6 Addressing Architecture, RFC 2373, July 1998.

[RFC2461] T. Narten, E. Nordmark, and W. Simpson, Neighbor Discovery for IP Version 6 (IPv6), RFC 2461, December 1998.

[RFC2529] B. Carpenter and C. Jung, Transmission of IPv6 over IPv4 Domains without Explicit Tunnels, RFC 2529, March 1999.

[RFC2663] P. Srisuresh and M. Holdrege, IP Network Address Translator (NAT) Terminology and Considerations, RFC 2663, August 1999.

[RFC2765] E. Nordmark, Stateless IP/ICMP Translation Algorithm (SIIT), RFC 2765, February 2000.

[RFC2766] G. Tsirtsis and P. Srisuresh, Network Address Translation–Protocol Translation (NAT-PT), RFC 2766, February 2000.

[RFC2767] K. Tsuchiya, H. Higuchi, and Y. Atarashi, Dual Stack Hosts Using the "Bump-in-the-Stack" Technique (BIS), RFC 2767, February 2000.

[RFC2893] R. Gilligan and E. Nordmark, Transition Mechanisms for IPv6 Hosts and Routers, RFC 2893, August 2000.

[RFC3056] B. Carpenter and K. Moore, Connection of IPv6 Domains via IPv4 Clouds, RFC 3056, February 2001.

[RFC3068] C. Huitema, An Anycast Prefix for 6to4 Relay Routers, RFC 3068, June 2001.

[RFC3142] J. Hagino and K. Yamamoto, An IPv6-to-IPv4 Transport Relay Translator, RFC 3142, June 2001.

[RFC3171] Z. Albanna, K. Almeroth, D. Meyer, M. Schipper, IANA Guidelines for IPv4 Multicast Address Assignment, RFC 3171, August 2001.

[RFC3338] S. Lee, M.-K. Shin, Y.-J. Kim, E. Nordmark, and A. Durand, Dual Stack Hosts Using "Bump-in-the-API" (BIA), RFC 3338, October 2002.

[RFC3489] J. Rosenberg, J. Weinberger, C. Huitema, and R. Mahy, STUN — Simple Traversal of User Datagram Protocol (UDP) through Network Address Translators (NATs), RFC 3489, March 2003.

[RFC4214] F. Templin, T. Gleeson, K. Talwar, D. Thaler, Intra-Site Automatic Tunnel Addressing Protocol (ISATAP), RFC 4214, October 2005.

[RFC4380] C. Huitema, Teredo: Tunneling IPv6 over UDP through Network Address Translations (NATs), RFC 4380, February 2006.

Chapter 7

IPv6 Network Software and Hardware

7.1 Introduction

In Chapter 6 several available mechanisms to support a transition to Internet Protocol version 6 (IPv6) were discussed. Frankly, this transition has not yet gathered steam as of press time in spite of the fact that a lot of the "science," the equipment, and some requirement (e.g., government mandate for U.S. Department of Defense [DoD]) has existed for up to a decade. The issue of address availability has temporarily been solved with the heavy use of network address translators (NATs). However, this use of NAT limits applications such as end-to-end public Voice-over-IP (VoIP), global IP-based cellular service, large sensor networks, and true IP addressability of billions of appliances, ranging from automobiles, personal digital assistants (PDAs), home appliances, and so on. As was implied in the previous chapters, the migration of IPv4 to IPv6 will not happen instantaneously; the expectation is that there will be a period of transition when both protocols are in use over the same infrastructure. As noted, to deal with this transition period, the designers of IPv6 have created technologies and address types so that IPv6 nodes can communicate with each other in a mixed environment, even if they are supported at the core by an IPv4-only infrastructure [MIC200701].

Organizations have been attempting to develop business cases that would show some short-term financial benefit for a conversion to IPv6. Such business cases may be difficult to develop in the immediate future, reminiscent of the desire for a business case to validate a move to VoIP in the late 1990s. However, at this juncture it is clear that VoIP is less expensive compared to traditionally priced telephony with built-in 20th century sociophilosophical pricing fee structures[1] and may be less costly when one considers the possibility of unified messaging and computer-telephony integration.

The introduction of IPv6 may need to parallel an even more basic transition: the introduction of personal computer (PC)-based automation in the mid-1980s. It was difficult to anticipate the eventual macroeconomic productivity gains that were finally realized in the late 1990s and early 2000s when this technology was first introduced. It took over a decade for the intrinsic and synergistic value of the deployment of PC/server technology to be manifestly and explicitly evident.

The expectation of observers is that, at some point in time, the technology will simply have self-initiated pull as more and more vendors bring IPv6 to the market. The transition may occur without much fanfare, initially on a case-by-case, application-by-application, island-by-island basis, until such time when critical mass is achieved.

The macro-level advantages of IPv6 will be evident perhaps a decade out from when the technology made its first deployment appearance. If the "technology pull" we mentioned will occur around 2010, then the full value of IPv6 will be more fully evident around 2020.

Given these observations, in this chapter we look at some of the practical considerations that may foster migration at some point in the future. Protocol transitions are never trivial, and the transition from IPv4 to IPv6 is particularly complex. Protocol transitions are typically achieved by installing and configuring the new protocol on all nodes within the network. Although this might be possible in a small or medium-size organization, the challenge of making a rapid protocol transition in a large organization or a carrier of any size (including an Internet Service Provider) is challenging. Therefore, although migration is the aspirational goal, consideration must be given to the short-term coexistence of IPv4 and IPv6 nodes. This can be achieved by installing and configuring the new protocol on a subset of nodes within the network.

7.2 IPv6 End Systems Applications

It has been a long-standing practice to separate the application-level functionality from the networking-level functionality. A typical business application is designed to run on a protocol stack via a service access point (SAP) interface (loosely, an application programming interface [API]) that isolates the business logic from the communication logic (see Figure 7.1). This implies that, in most instances, existing applications should be able to operate equally well on an IPv4 stack as on an IPv6 stack. The only exception may be applications that have a rather "intimate" association with a (large) population of remote endpoint devices.

---

[1] For example, this idea is that "long distance service" is a "luxury" item, and therefore people making long-distance calls should subsidize people making local calls when, in fact, the economics are exactly the opposite — the implication of our statement is that VoIP may or may not actually be cheaper in a true-cost comparison, but it is cheaper in a price comparison.

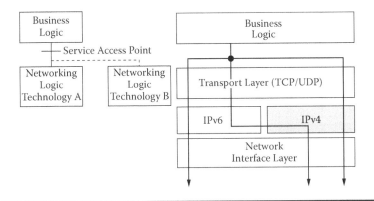

Figure 7.1 Well-designed business application.

7.3 IPv6 End Systems Communications Software

Considerable progress has been made in the recent past in enabling clients (e.g., PCs) to be IPv6 enabled. For example, Linux- and Microsoft-based systems support IPv6. This section focuses on the announced support of IPv6 by the Microsoft® Windows® and Windows Server® operating systems.

Both the Microsoft® Windows Vista™ operating system and the Windows Server® operating system, code name Longhorn (in beta testing at press time), include the Next Generation Transmission Control Protocol/Internet Protocol (TCP/IP) stack, a redesigned TCP/IP stack with an integrated version of both IPv4 and IPv6 [MIC200601]. IPv6 support in the Windows Vista™ and Windows Server® Longhorn operating systems includes the following:

- Dual IP layer architecture
- Support is installed and enabled by default
- Graphical user interface (GUI)-based configuration
- Full support for Internet Protocol security (IPsec)
- Multicast Listener Discovery version 2 (MLDv2)
- Link-local multicast name resolution (LLMNR)
- Literal IPv6 addresses in universal resource locators (URLs)
- Support for ipv6-literal.net names
- IPv6 over Point-to-Point Protocol (PPP)
- Dynamic Host Control Protocol for version 6 (DHCPv6)
- Random interface IDs
  - Updates to Teredo
  - Enhanced security for IPv6 and Teredo
  - Features to disable IPV6 components

The information that follows in this subsection is based on [MIC200601] and is included to provide an example of the available IPv6 technology.

7.3.1 Dual-IP-Layer Architecture

The implementation of IPv6 in the Windows® XP and Windows Server® 2003 operating systems is a dual-stack architecture that has separate protocol components for IPv4 and IPv6 that are installed through the network connections folder. The separate IPv4 and IPv6 components each had their own transport layer that included TCP and User Datagram Protocol (UDP) and framing layer. The Next Generation TCP/IP stack is a single protocol component installed through the network connections folder that supports the dual-IP-layer architecture, in which both IPv4 and IPv6 share common transport and framing layers.

Because there is a single implementation of TCP, TCP traffic over IPv6 can take advantage of all the performance features of the Next Generation TCP/IP stack. These features include all of the performance enhancements of the IPv4 stack of the Windows® XP and Windows Server® 2003 operating systems and additional enhancements new to the Next Generation TCP/IP stack, such as receive window auto tuning and compound TCP — which can dramatically improve performance on high-latency/high-delay connections — and better support for TCP traffic in high-loss environments (such as wireless local area networks [LANs]).

7.3.2 Installed and Enabled by Default

In the Windows Vista™ and Windows Server® Longhorn operating systems, IPv6 is installed and enabled by default as the Internet Protocol version 6 (TCP/IPv6) component from the properties of a connection in the connections and adapters folder. In the Windows Vista™ and Windows Server® Longhorn operating systems, many operating system components now support IPv6.

When both IPv4 and IPv6 are enabled, the Next Generation TCP/IP stack prefers the use of IPv6. For example, if a Domain Name System (DNS) name query response message contains a list of both IPv6 and IPv4 addresses, the Next Generation TCP/IP stack will attempt to communicate over IPv6 first, subject to the address selection rules that are defined in RFC 3484.

The preference of IPv6 over IPv4 offers IPv6-enabled applications better network connectivity because IPv6 connections can use IPv6 transition technologies such as Teredo, which allow peer or server applications to operate behind NATs without requiring NAT configuration or application modification.

Enabling IPv6 by default and preferring of IPv6 traffic does not impair IPv4 connectivity. For example, on networks without IPv6 records in the DNS infrastructure, communications using IPv6 addresses are not attempted unless the user or application specifies the destination IPv6 address.

To take advantage of IPv6 connectivity, networking applications must be updated to use Windows®-based sockets functions that are not specific to IPv4 or IPv6.

7.3.3 GUI-Based Configuration

In the Windows® XP and Windows Server® 2003 operating systems, one must manually configure IPv6 configuration settings with netsh interface ipv6 commands at a Windows command prompt. Windows Vista™ and Windows Server® Longhorn operating systems now allow you to also manually configure IPv6 settings through the properties of the Internet Protocol version 6 (TCP/IPv6) component in the connections folder.

7.3.4 Full Support for IPsec

Internet Protocol security (IPsec) support for IPv6 traffic in the Windows® XP and Windows Server® 2003 operating systems is limited. There is no support for Internet key exchange (IKE) or data encryption. IPsec security policies, security associations, and keys are configured through text files and activated through a command line tool, IPsec6.exe.

In the Windows Vista™ and Windows Server® Longhorn operating systems, IPsec support for IPv6 traffic is the same as that for IPv4, including support for IKE and data encryption with Advanced Encryption Standard (AES) 128/192/256. The IP Security Policies snap-in now supports the configuration of IPsec policies for IPv6 traffic in the same way as IPv4 traffic using either the IP Security Policies snap-in or the new Windows® Firewall with Advanced Security snap-in. See Chapter 10 for more information on IPsec.

7.3.5 Multicast Listener Discovery Version 2 (MLDv2)

Windows Vista™ and Windows Server® Longhorn operating systems support MLDv2, specified in RFC 3810, which allows IPv6 hosts to register interest in source-specific multicast traffic with their local multicast routers. A host running on the Windows Vista™ or Windows Server® Longhorn operating systems can register interest in receiving IPv6 multicast traffic from only specific source addresses (an include list) or from any source except specific source addresses (an exclude list).

7.3.6 Link-Local Multicast Name Resolution (LLMNR)

Windows Vista™ and Windows Server® Longhorn operating systems support LLMNR, which allows IPv6 hosts on a single subnet without a DNS server to resolve each other's names. This capability is useful for single-subnet home networks and ad hoc wireless networks. Rather than unicasting a DNS query to a DNS server, LLMNR nodes send their DNS queries to a multicast address on which all the

LLMNR-capable nodes of the subnet are listening. The owner of the queried name sends a unicast response. IPv4 nodes can also use LLMNR to perform local subnet name resolution without having to rely on NetBIOS over TCP/IP broadcasts.

7.3.7 Literal IPv6 Addresses in URLs

The WinINet API in the Windows Vista™ and Windows Server® Longhorn operating systems supports RFC 2732 and the use of IPv6 literal addresses in URLs. For example, to connect to the Web server at the IPv6 address 2001:DB8:100:2A5F::1, a user with a WinINet-based Web browser (such as Internet Explorer) can type http://[2001:DB8:100:2A5F::1] as the URL. Although typical users might not use IPv6 literal addresses, the ability to specify the IPv6 address in the URL is valuable to application developers, software testers, and network troubleshooters.

7.3.8 Support for ipv6-literal.net Names

Windows Vista™ and Windows Server® Longhorn operating systems support the use of *IPv6Address*.ipv6-literal.net names. The *IPv6Address*.ipv6-literal.net name resolves to *IPv6Address*. For example, the 2001:DB8:28:3:F98A:5B31:67B7:67EF. ipv6-literal.net name resolves to 2001:DB8:28:3:F98A:5B31:67B7:67EF. The IPv6 address in the name can be global, unique local, or link local (with or without a zone ID). To specify a zone ID (also known as a scope ID), replace with an "s" the "%" used to separate the IPv6 address from the zone ID. For example, to specify the destination FE80::218:8BFF:FE17:A226%4, the name is FE80::218:8BFF:FE17: A226s4.ipv6-literal.net.

An ipv6-literal.net name can be used in services or applications that do not recognize the syntax of normal IPv6 addresses. It is always preferable to use a DNS name that corresponds to a destination, such as filesrv1.example.com. However, the ipv6-literal.net name can be used for connectivity when the DNS name for the destination is not registered, and the IPv6 address is known.

To use an ipv6-literal.net name in the computer name part of a Universal Naming Convention (UNC) path, convert the colons (:) in the address to dashes (-). For example, to specify the Docs share of the computer with the IPv6 address of 2001:DB8:28:3:F98A:5B31:67B7:67EF, use the UNC path \\2001-db8-28-3-f98a-5b31-67b7-67ef.ipv6-literal.net\docs.

The ipv6-literal.net name is an alias for the name of the file server.

7.3.9 IPv6 over PPP

The built-in remote access client now supports the IPv6 Control Protocol (IPV6CP), as defined in [RFC2472], to configure IPv6 nodes on a PPP link. Native IPv6 traffic can now be sent over PPP-based connections. For example, IPV6CP support allows you to connect with an IPv6-based Internet Service Provider (ISP) through dial-up or PPP over Ethernet (PPPoE)-based connections that might be used for broadband Internet access. In addition, IPV6CP supports Layer Two Tunneling Protocol (L2TP)-based virtual private network connections.

7.3.10 Dynamic Host Control Protocol Version 6

The DHCP Client service in Windows Vista™ and Windows Server® Longhorn operating systems supports DHCPv6 defined in RFCs 3315 and 3736. A computer running Windows Vista™ or Windows Server® Longhorn operating systems can perform both DHCPv6 stateful and stateless configuration on a native IPv6 network.

7.3.11 Random Interface IDs

To prevent address scans of IPv6 addresses based on the known company IDs of network adapter manufacturers, Windows Vista™ and Windows Server® Longhorn operating systems by default generate random interface IDs for nontemporary autoconfigured IPv6 addresses, including public and link-local addresses. A public IPv6 address is a global address that is registered in DNS and is typically used by server applications for incoming connections, such as a Web server.

Note that this new behavior is different from that for temporary IPv6 addresses, as described in RFC 3041. Temporary addresses also use randomly derived interface IDs. However, they are not registered in DNS and are typically used by client applications when initiating communication, such as a Web browser.

One can disable this behavior with the **netsh interface ipv6 set global rando mizeidentifiers=disabled** command. You can enable this behavior with the **netsh interface ipv6 set global randomizeidentifiers=enabled** command.

7.3.12 Updates to Teredo

As discussed in Chapter 6, Teredo is an IPv6 transition technology that allows IPv6/IPv4 nodes that are separated by one or more NATs to communicate end to end with global IPv6 addresses. NATs are commonly used on the Internet to preserve the public IPv4 address space by translating the addresses and port numbers of traffic to and from private network hosts that use private IPv4 addresses.

Although NATs extend the life of the public IPv4 address space, this functionality comes at the cost of violating the original design principle of the Internet that all nodes should communicate with a unique global address. Because of the reuse of private addresses and the translation between private and public addresses that occur at the NAT, servers and peers that are located on private networks behind NATs cannot communicate without either manually configuring the NAT or modifying application protocols.

Although IPv4 traffic for servers and peers that are behind an NAT might have problems traversing an NAT, Teredo-based IPv6 traffic can traverse an NAT without having to configure the NAT or modify application protocols. Teredo IPv6 addresses are global addresses, unique to the entire Internet. Teredo restores global addressing and end-to-end connectivity for IPv6 traffic for an environment that does not support global addressing and end-to-end connectivity for IPv4 traffic.

Teredo was first released with the Advanced Networking Pack for Windows® XP with Service Pack 1 and is included with Windows® XP Service Pack 2 and Windows Server® 2003 Service Pack 1. Windows Vista™ and Windows Server® Longhorn operating systems also support Teredo.

In the Windows Vista™ operating system, Teredo is enabled but inactive by default. To become active, you must either use an application that requires Teredo or configure advanced settings on a Windows® Firewall inbound rule to allow edge traversal. In the Windows Server® Longhorn operating system, Teredo is disabled by default.

Teredo in Windows Vista™ and Windows Server® Longhorn operating systems supports the following new features:

- Teredo is now enabled for domain member computers. Teredo for Windows® XP and Windows Server® 2003 operating systems automatically disabled itself if the computer was a member of a domain. A domain member computer is more likely to be attached to a network that has deployed either native IPv6 connectivity or Intrasite Automatic Tunnel Addressing Protocol (ISATAP), an IPv6 transition technology. However, domain member computers can also benefit from Teredo-based IPv6 connectivity.
- Teredo can now work if there is one Teredo client behind one or more symmetric NATs. (See Section 6.4.6.1 for a detailed discussion of the various NAT types.) A symmetric NAT maps the same internal (private) address and port number to different external (public) addresses and ports, depending on the external destination address (for outbound traffic). For example, Teredo in Windows Vista™ and Windows Server® Longhorn operating systems will work if one of the peers is behind a symmetric NAT and the other is behind a cone or restricted NAT. Teredo for Windows® XP and Windows Server® 2003 operating systems disables itself if it detects that it is behind a symmetric NAT. This new behavior allows Teredo to work between a larger set of Internet-connected hosts.
- The Windows Vista™ operating system now has support for UPnP™-certified symmetric NATs. If one has connectivity problems, one can enable UPnP on your symmetric NAT for improvements in connectivity.

7.3.13 Enhanced Security for IPv6 and Teredo

Having IPv6 and Teredo enabled by default does not make one's computer more vulnerable to attack by malicious users or programs because of the following:

■ Windows® Firewall, included with and enabled by default for both the Windows Vista™ and Windows Server® Longhorn operating systems, is a stateful host-based firewall for both IPv4 and IPv6 traffic. All of the protections against unwanted, unsolicited, incoming traffic apply to both IPv4 and IPv6 traffic.

■ Windows® Firewall allows exceptions for wanted, unsolicited, incoming traffic based on TCP or UDP ports or by specifying a program name and applying it to an individual computer. Windows® Firewall-based exceptions are much more specific than exceptions configured on typical NATs.

■ The Windows® Filtering Platform is a new architecture in the Windows Vista™ and Windows Server® Longhorn operating systems that allows third-party software developers access to the TCP/IP packet-processing path, in which outgoing and incoming packets can be examined or changed before allowing them to be processed further. By tapping into the TCP/IP processing path, independent software vendors (ISVs) can create firewalls, antivirus software, diagnostic software, and other types of applications and services. The Windows® Filtering Platform is designed for both IPv4 and IPv6 traffic. Third-party host-based firewall products that use the Windows® Filtering Platform will typically support both IPv4 and IPv6 traffic.

Computers running the Windows Vista™ operating system have IPv6, Teredo, and Windows® Firewall enabled by default and are protected from unwanted, unsolicited, incoming IPv6 traffic.

7.3.14 Features to Disable IPv6 Components

Unlike the Windows® XP operating system, IPv6 in the Windows Vista™ and Windows Server® Longhorn operating systems cannot be uninstalled. To disable IPv6 on a specific connection, one can do the following:

■ In the Network connections folder, obtain properties of the connection and clear the check box next to the Internet Protocol version 6 (TCP/IPv6) component in the list under "This connection uses the following items." This method disables IPv6 on your LAN interfaces and connections but does not disable IPv6 on tunnel interfaces or the IPv6 loopback interface.

To selectively disable IPv6 components and configure behaviors for IPv6 in the Windows Vista™ operating system, create and configure the following registry value (DWORD type):

■ HKEY_LOCAL_MACHINE\SYSTEM\CurrentControlSet\Services\tcpip6\ Parameters\DisabledComponents

DisabledComponents is set to 0 by default.

The DisabledComponents registry value is a bit mask that controls the following series of flags, starting with the low order bit (Bit 0):

- Bit 0. Set to 1 to disable all IPv6 tunnel interfaces, including ISATAP, 6to4, and Teredo tunnels. Default value is 0.
- Bit 1. Set to 1 to disable all 6to4-based interfaces. Default value is 0.
- Bit 2. Set to 1 to disable all ISATAP-based interfaces. Default value is 0.
- Bit 3. Set to 1 to disable all Teredo-based interfaces. Default value is 0.
- Bit 4. Set to 1 to disable IPv6 over all nontunnel interfaces, including LAN interfaces and PPP-based interfaces. Default value is 0.
- Bit 5. Set to 1 to modify the default prefix policy table to prefer IPv4 to IPv6 when attempting connections. Default value is 0.

To determine the value of DisabledComponents for a specific set of bits, construct a binary number consisting of the bits and their values in their correct position and convert the resulting number to hexadecimal. For example, if you want to disable 6to4 interfaces, disable Teredo interfaces, and prefer IPv4 to IPv6, you would construct the following binary number: 101010. When converted to hexadecimal, the value of DisabledComponents is 0x2A.

The following table lists some common configuration combinations and the corresponding value of DisabledComponents.

Configuration Combination	DisabledComponents Value
Disable all tunnel interfaces	0x1
Disable 6to4	0x2
Disable ISATAP	0x4
Disable Teredo	0x8
Disable Teredo and 6to4	0xA
Disable all LAN and PPP interfaces	0x10
Disable all LAN, PPP, and tunnel interfaces	0x11
Prefer IPv4 over IPv6	0x20
Disable IPv6 over all interfaces and prefer IPv4 to IPv6	0xFF

One must restart the computer for the changes to the DisabledComponents registry value to take effect.

7.4 IPv6 Support by Major Router Vendors

Major router manufacturers have supported IPv6 for a number of years. This section highlights that support and is meant to give a sample of the level of support for IPv6, and focuses only on the top market leaders.

7.4.1 Cisco Systems

Cisco Systems implements the main IPv6 transition mechanisms — dual stack, tunneling and translation — as part of its IPv6 solution. Cisco has been active in the definition and implementation of the IPv6 architecture within the Internet Engineering Task Force (IETF) IPv6 standardization effort, holding co-chair positions in multiple IETF Working Groups. Cisco was also a founding member of the IPv6 Forum.

The Cisco IOS operating system is the software used on the vast majority of Cisco Systems routers. Cisco IOS releases are split into several categories, termed *trains*, each containing a different set of features. Trains generally map onto distinct markets or groups of customers that Cisco is targeting.

The mainline train is designed to be the most stable release the company can offer, and its feature set never expands during its lifetime. Updates are released only to address bugs in the product.

- The T, technology, train gets new features and bug fixes throughout its life and is therefore less stable than the mainline.
- The S, service provider, train is heavily customized for service provider customers.
- The E, enterprise, train is customized for implementation in enterprise environments.
- The B, broadband, train supports Internet-based broadband features.

In 2000, Cisco announced an IPv6 roadmap for its IOS operating system. In 2001, Cisco IOS Software Release 12.2T incorporated IPv6 in its first commercial release. Subsequently, Cisco Software Release 12.0S enabled the support of IPv6 in core service provider infrastructures. Currently Cisco IOS releases enable IPv6 in a wide range of Cisco products. Major Cisco router products, the IOS release, and the product use are shown in the following table.

Product	IOS Release	Comments
Cisco 7600 Series	IOS Software Release 12.2SR	High-end service provider edge network device IPv6 hardware forwarding assistance
Cisco 12000 Series	IOS Software Release 12.0S	Service provider infrastructure IPv6 hardware forwarding assistance
Cisco 12000 Series	IOS-XR Software Release	High-performance and high-availability core network applications IPv6 hardware forwarding assistance
Cisco CRS-1 Carrier Routing System	IOS-XR Software Release	High-performance and high-availability core network applications IPv6 hardware forwarding assistance

A full range of IPv6 features is supported in the various releases of IOS, including IPv6 unicast address types, ICMPv6, IPv6 Neighbor Discovery, and IPv6 MTU (maximum transfer unit) Path Discovery. Some features that are required only on edge routers are implemented in IOS software releases targeted at these devices. For example, IPv4-compatible tunneling and ISATAP tunneling are available on the 12.2SR release but not the 12.0S release.

7.4.2 Juniper Networks

Juniper Networks is a supplier of core and edge routers in service provider networks. Major products include the T-Series, focusing on service provider core requirement, and the M-Series, focusing on edge requirements. Hardware acceleration of IPv6 forwarding is available in both series. Both the T- and M-Series of routers use the JUNOS purpose-built Internet software operating system across the entire product line. JUNOS has supported a wide range of IPv6 features since Release 5.1 in November 2001. It has received the IPv6 Ready certification from the IPv6 Forum, indicating that the product has met basic implementation and interoperability requirements for the core IPv6 specifications.

Fundamental IPv6 features include the following:

■ Support of IPv6 address types
■ Stateless autoconfiguration
■ Neighbor discovery
■ ICMPv6
■ A number of IPv4-to-IPv6 transition mechanisms, including dual stack and configured tunnels

7.4.3 Alcatel-Lucent

Alcatel-Lucent is a supplier of edge routers in the service provider market with its 7750 and 7710 Service Router (SR) products. The Alcatel-Lucent 7750 SR is a multi-service edge router purpose built for service providers who are delivering next-generation business and residential services. The 7750 SR is specifically engineered to deliver IP services like Layer 2 and Layer 3 virtual private networks (VPNs) over an IP/MPLS (Multi-Protocol Label Switching) backbone. The 7710 SR is designed for smaller POPs (points of presence) and is also a solution for enterprise customers needing high availability

The Service Router Operating System (SR-OS) is an integral component of the SR family and is used on both the 7750 SR and the 7710 SR. The SR-OS supports a number of IPv6 features including the following:

■ IPv6 address architecture
■ Stateless address autoconfiguration

- Neighbor discovery
- ICMPv6
- OSPF (Open Shortest Path First) and BGP-4 (Border Gateway Protocol 4) routing extensions for IPv6
- 6over4 transition mechanism

The SR-OS has been certified with the IP Ready Logo from the University of New Hampshire InterOperability Laboratory (UNH-IOL).

7.5 Carrier/ISP IPv6 Services

This section presents a sample of IPv6-related services available in Asia-Pacific, Europe, and North America. With access to almost 70 percent of the IPv4 address space, IP address shortages have not been a major issue in the North American community. Consequently, the number of IPv6-related testbeds, experiments, service offerings, and accordingly the level of widespread experience with IPv6 in North America pales in contrast to the support and experience found in places like Japan, Korea, and Europe.

7.5.1 Asia Pacific

7.5.1.1 NTT Communications IPv6-Related Services

7.5.1.1.1 IPv6 Gateway Service

The IPv6 Gateway Service provides native connectivity to NTT's global backbone jointly operated by Verio. Maximum bandwidth is 155 Mbps. The same service is offered in Japan, the United States, Europe, Korea, Taiwan, Hong Kong, Australia, Malaysia, and other countries.

7.5.1.1.2 IPv6 Tunneling Service (IPv6 over IPv4 tunneling)

The IPv6 Tunneling Service (IPv6 over IPv4 tunneling) provides an IPv6 connectivity service for NTT users via IPv6 over IPv4 tunnels. It is offered through a variety of access technologies, including dedicated line, Integrated Services Digital Network (ISDN), Asymmetrical Digital Subscriber Line (ADSL), and optical. Verio, a subsidiary of NTT Communications, has expanded the reach of the IPv6 Gateway Services to a larger user base by making the commercial offering available through any U.S.-based Internet connection, regardless of the Internet access provider. With the use of a router capable of supporting IPv6 traffic, customers without a direct connection to the NTT Communications Global IP Network can now receive IPv6 Tunneling Service. This off-net configured static IPv6 tunnel

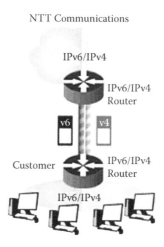

Figure 7.2   NNT Communications IPv4/IPv6 Dual Service.

offering provides an easy way to receive IPv6 connectivity for testing and development purposes.

7.5.1.1.3   ADSL IPv6 Dual Service

The ADSL IPv6 Dual Service is an IPv4/IPv6 dual-connectivity service through Assymetrical Digital Subscriber Line (ADSL) access, shown in Figure 7.2. Network prefixes are automatically assigned to consumer premise equipment (CPE) routers.

7.5.1.1.4   Multicast Service

Multicast Service is a multicast service for video distribution targeted for ISPs and enterprises. It supports both IPv4 and IPv6. This multicast service enables real-time video streaming to many users with minimal burden on the video server.

7.5.1.1.5   Housing Connectivity Service IPv6/IPv4 Dual Stack

The Housing Connectivity Service IPv6/IPv4 Dual Stack service offers dual IPv4/IPv6 stack direct Ethernet connectivity from the customer servers housed in NTT's data centers.

7.5.1.1.6   Super IPv6 Dual Ether Access

The Super IPv6 Dual Ether Access service adds IPv6 connectivity via an Ethernet interface to customer sites for direct connectivity to NTT's backbone.

7.5.1.2 *KDDI/KDDI Lab*

7.5.1.2.1 KDDI Internet IPv6 Native Service

Started as commercial service in November 2002, KDDI Internet IPv6 Native Service provides native IPv6 connectivity through dedicated line and Asynchronous Transfer Mode (ATM) access line services.

7.5.1.2.2 IPv6 Tunneling Service

IPv6 Tunneling Service is offered to users with a fixed IPv4 address option on KDDI's Internet service with dedicated line, Ethernet, or Digital Subscriber Line (DSL) access. IPv6 address allocation is /48 for end users and /43 for other ISPs.

7.5.1.2.3 6to4

KDDI Lab is operating a 6to4 relay router. Any users with global IPv4 addresses can gain IPv6 connectivity through tunneling by this 6to4 relay router without registration.

7.5.1.3 *Japan Telecom*

7.5.1.3.1 IPv6 Tunneling Service

The IPv6 Tunneling Service is a commercial IPv6 service for its business users and is available for a wide range of access technologies, from dedicated line to wireless.

7.5.1.3.2 Native Service

Native Service offers IPv6-only connectivity over Ethernet as part of its connectivity services for businesses.

7.5.1.3.3 Dual Stack Service

Dual-Stack Service offers IPv4 and IPv6 connectivity over Ethernet as part of its connectivity services for businesses.

7.5.2 *Europe*

7.5.2.1 *European Internet Exchange Association*

There are many thousands of ISP networks in Europe, and many of these establish peering relationships at Internet exchange points (IXPs). It would not be cost-effective, scalable, or manageable to interconnect with all of them individually. An

IXP is a single physical network infrastructure (often an Ethernet LAN) to which many ISPs can connect. Any ISP that is connected to the IXP can exchange traffic with any of the other ISPs connected to the IXP, using a single physical connection to the IXP, thus overcoming the scalability problem of individual interconnections. Practically all of the European IXPs support IPv6 for ISP interconnection.

Note that IXPs are not generally involved in the peering agreements between connected ISPs, but do however have requirements that an ISP must meet to connect to the IXP.

7.5.2.2 BT — UK6x

UK6x is an IPv6 Internet exchange operated by BT Exact. The UK6x is located in London and offers a set of IPv6 network and application services including the following:

- **Native IPv6 connections:** Direct cabling to the UK6x using Ethernet, Optical/ATM, and so on.
- **Tunneled IPv6 connections:** Using a variety of remote access services, including dedicated line and dial.
- **Peering:** Open peering facilities are provided to connect to other networks.
- **Transit:** IPv6 transit services are provided to connect to the Internet backbone.
- **Address allocation:** IPv6 addresses are allocated in blocks ranging from individual to national ISP size.
- **IPv4-IPv6 interworking:** Mechanisms to facilitate interworking between IPv4 and IPv6 are provided.

7.5.3 North America

7.5.3.1 Moonv6

The North American IP community's interest in IPv6 has been somewhat lacking in comparison to that of Europe and Asia. To rectify this shortcoming, the North American IPv6 Task Force, in collaboration with the UNH-IOL, the Joint Interoperability Test Command (JITC), and the U.S. DoD, have sponsored the Moonv6 project. Taking place across the United States at multiple locations, the Moonv6 project is a large, permanently deployed, multi-vendor IPv6 network and plays a significant role in ensuring interoperability and migration objectives are identified and demonstrated.

7.5.3.2 AT&T

AT&T is a participant in the Moonv6 project just described. Through connectivity to the North American Task Force's Moonv6 next-generation Internet network,

AT&T offers access to the broader IPv6 Internet using either tunneling technologies or an overlay network for native IPv6 support. AT&T has also established IPv6 peerings (network interconnections) with other IP backbone providers, such as Global Crossing, and allows seamless exchange of IPv6 traffic across the networks.

AT&T also offers a range of services to support the U.S. federal government's IPv6 mandate. These include the following:

- IPv6 Internet Connectivity Services providing connectivity to the IPv6 Internet over multiple access technologies (PPP, MultiLink PPP [MLPPP], Frame Relay, and ATM) for customer access, typically from large agency locations.
- Remote Access Service to IPv6 Internet for small (or satellite) locations and individual remote users. Users access a dynamically configurable IPv6 tunnel gateway through IPv4 ISPs using fractional T1, DSL, or dial-up access. The Tunnel Setup Protocol (TSP) is used to create tunnels to transport IPv6 traffic over an IPv4 network to the gateway.

7.5.3.3 Global Crossing

Global Crossing has IPv6 natively installed on all its IP services access routers. IPv6 is enabled on the IP-based transport services provided by Global Crossing's IP backbones under the product names of Dedicated Internet Access and IP VPN. Global Crossing's Dedicated Internet Access is a scalable Internet access service provisioned directly onto the Global Crossing fiber-based network backbone. It is available as fixed/full-pipe, variable/tiered, or variable/burstable service, from 64 kbps service through SONET/SDH OC-48/STM-16 and 10B-E, FastE, Gigabit Ethernet (GigE).

IPv6 has been generally available for Internet Services customers since mid-2005. Global Crossing has also implemented a number of IPv6 peering relationships with partners, and its customer base includes a number of carriers in various regions of the globe. Dual-stack access services are provided to assist in customers' transition to IPv6. Other IPv6-related features include native IPv4/IPv6 over MPLS, IPv6 addresses, and IPv6 DNS.

7.6 IPv6 in Wireless Networks

Because mobile networks are a major driver in the transition to IPv6, an overview of the major IPv6 activities affecting wireless carriers is presented in this section. The wireless mobile market is perceived as one of the main factors for IPv6 adoption in Europe and Asia. Wireless data services are expected to encourage the adoption of IPv6 because of the quickly expanding number of IP-addressable devices. These services are expected to increase with the move to the Third-Generation Partnership Program Universal Mobile Telephony System Technical Specifications

(3GPP UMTS), including the adoption of IP Multimedia Services (IMS) to allow the addition of real-time IP-based services. IMS will be the most important factor for the adoption of IPv6 within mobile operator networks. IMS is based on three primary technologies:

■ SIP (Session Initiation Protocol) as the control plane for the establishment of IMS sessions
■ IPv6
■ Diameter for authentication, authorization, accounting (AAA), and billing.

Smooth transition to IPv6 in IP backbones, and possibly later in the Radio Access Network (RAN), is the goal stated in 3GPP specifications.

IPv6 had been originally mandated for IMS in UMTS Release 5. Since then, IPv4-based user equipment (UE) and IMS dual-stack user equipment have been introduced in the IMS specifications. Internetworking and migration scenarios between IPv4-based IMS and IPv6-based IMS are ongoing working items that should be completed as part of 3GPP Release 6.

The IETF has also been active in proposing the use of IPv6 in wireless networks:

■ Soininen [SOI200301] describes different scenarios in 3GPP packet networks that would need IPv6 and IPv4 transition. The focus of this document is on scenarios in which the UE connects to nodes in other networks (e.g., in the Internet). The purpose of the document is to list the scenarios for further discussion and study.
■ Wiljakka [WIL200501] analyzes the transition to IPv6 in 3GPP packet networks. The focus is on analyzing different transition scenarios and applicable transition mechanisms and finding solutions for those transition scenarios.

References

[MIC200601] Microsoft Corporation Materials on Vista.
[MIC200701] Microsoft Corporation, IPv6 Transition Technologies, White Paper, October 2003, updated January 2007.
[RFC2472] D. Haskin, E. Allen, IP Version 6 over PPP, RFC 2472, December 1998.
[SOI200301] J. Soininen, Transition Scenarios for 3GPP Networks, RFC 3574, August 2003.
[WIL200501] J. Wiljakka, Analysis on IPv6 Transition in Third Generation Partnership Project (3GPP) Networks, RFC 4215, October 2005.

Chapter 8

Implementing IPv6 Transition Strategies

8.1 Introduction

This chapter provides example steps required to effect a transition to Internet Protocol version 6 (IPv6) in an actual networking environment. Although the chapter focuses on activities typically followed by Internet Service Providers (ISPs), the general approach is applicable to other types of networks, including corporate intranets and institutional networks. This material is based largely on work by the 6NET organization in support of the European National Research and Education Networks (NRENs) [6NE200501]. The 6NET project has involved the participation of some 15 NRENs. The NRENs provide connectivity to universities directly, or to metropolitan area networks (MANs), which then connect the universities. The NRENs have existing production IPv4 networks, the efficient and reliable running of which are paramount to their operation. Methods of introducing IPv6 into these networks are required that both enable IPv6 at the same level of performance as IPv4 and do not adversely impact the performance of the production IPv4 network.

Related work has also been done by the Internet Engineering Task Force (IETF) IPv6 Operations Working Group (V6OPS), which has undertaken a study of IPv6 transition scenarios for ISPs [RFC4029]. This work is also summarized in this chapter. The scope of the IETF work is broader than that of the 6NET NREN due to the differing focus of the services deployed by NRENs and commercial ISPs; for example, NRENs do not generally deploy broadband services out to end users.

8.2 Summary of NREN Transition Recommendations

This section discusses potential transition mechanisms for the NRENs and the advantages and disadvantages of the mechanisms.

8.2.1 General Approach for NREN Transition

A general approach for an ISP looking to transition is given in the IETF V6OPS document [RFC4029] and in the case of a 6NET NREN, this approach may be broken down into broad stages as follows:

1. Obtain an IPv6 address space, most likely a /32 SubTLA (from the RIPE network control center (NCC) in Europe).
2. Devise an IPv6 address allocation plan for the NREN network and the end-site universities.
3. Study the available tools for network management and monitoring and establish any required new operational procedures.
4. Select the appropriate transition path for IPv6 transport over the NREN network infrastructure (as mentioned, this may include direct transition to dual stack, use of Multi-Protocol Label Switching (MPLS) or Asynchronous Transfer Mode (ATM) permanent virtual circuits (PVCs)).
5. Select the appropriate IPv6 routing protocol and decide the routing policy (which may be the same as IPv4).
6. Deploy any necessary transition aids (e.g., 6to4 relay router).
7. IPv6 enable any required services (e.g., Domain Name System (DNS), quality of service (QoS), or multicast).
8. Follow the best practice for secured transition mechanism deployment.
9. Apply Steps 2–8 for any regional networks attached to the NREN backbone between the NREN and end sites.
10. Enable the equipment in the end-site premises.

If a native dual-stack approach is not enabled initially, the interim method (e.g., MPLS) should be upgraded to IPv6 native when procurement allows. The specific transition path may depend on existing technology and equipment/software availability as well as nontechnical issues such as available budget.

The 6NET report also recommends that the resilience and robustness of the platform be tested in a test environment before production deployment.

Finally, 6NET reports on (transition) security issues in [6NET-D312] and [6NET-D622].

8.2.2 Dual-Stack Issues

When performing dual-stack transition at the NREN scope, the ultimate goal is to make the same production routers route and forward both IPv4 and IPv6 traffic (and thus not create any kind of virtual overlay network, e.g., via MPLS or ATM PVCs).

The steps in deploying a dual-stack IPv4-IPv6 network on common infrastructure may include:

- Create and operate a test network with IPv4-tunneled (or even MPLS or ATM) connections to gain perspective on the operation of IPv6.
- Evaluate the router software versions in the test environment to see if they are stable and robust enough to be used in the main network with IPv4 and IPv6 together and how IPv6 affects IPv4 performance.
- If they are stable, start upgrading production routers to IPv4/IPv6 and enable IPv6 on the links that are used. Usually, the network topology will be the same as with IPv4.
- If problems (e.g., severe bugs affecting production services) arise, either try to fix or avoid them or drop back to an IPv4-only operation.

As mentioned previously, some router vendors have been shipping IPv6 capability in their production software for some time, and this experience has enhanced the stability for core features. This is a big advantage for emerging commercial ISP deployments. To break the "chicken-and-egg" status, it is desirable to deploy IPv6 in advance of heavy demand as there will not be demand until the service is well supported by the ISPs. By introducing dual-stack networking throughout the core networks, provision and deployment issues are pushed to the edge, such that they become a per-site issue. In academic networks, a commercial case for deployment is not generally required because the service is usually provisioned to enable research.

The advantage of dual-stack operation is that the network is the same for IPv4 and IPv6: There need not be new routers for IPv6, and there is no need to maintain a potentially complex overlay network. Similarly, this is also a disadvantage; because the network is the same, the problems (especially software bugs), if such arose, could also affect IPv4 services, which would probably not be the case if the network was separated. One also has to be aware of the performance impact of running IPv6, especially if the dual-stack implementation is not in hardware, and IPv4 routers are encapsulating IPv6 in software.

8.2.3 General Tunneling Issues

There is an advantage to using IPv6 tunnels over the existing IPv4 infrastructure — namely, the infrastructure is already tuned to perform well. and thus even with the tunneling overhead, the IPv6 overlay should also perform well.

Where the tunnels are configured manually, it is quite possible that the tunnels do not always take an optimal path between sites, where one IPv6 hop may span many IPv4 hops. Automatic tunneling, for example, with 6to4 as described in Chapter 6, routes IPv6 traffic over IPv4 tunnels by the most efficient IPv4 path between two 6to4 gateways. The problem arises with the interaction with other (non-6to4) IPv6 gateways, for which the routing may be very unpredictable or even not available at all.

The dependence on the existing IPv4 infrastructure may be a weakness (e.g. software problems, denial of service attacks against routers, etc.) that would also affect the IPv6 service. That said, one would expect the production IPv4 service to be well supported, so such issues ought to be rare.

In IPv6, path maximum transfer unit (MTU) discovery and management, fragmentation, and reassembly are handled by the end hosts, not intermediate routers. This has a number of implications. For example, in Ethernet networks, the MTU is 1500. Using tunneling on the interfaces where the MTU is 1500 reduces the usual path MTU to 1480 bytes, which will add some latency as path MTU discovery is initiated.

8.2.4 IPv6 over MPLS Issues

MPLS is growing in popularity in the European NRENs, primarily for use in traffic engineering. There are two likely usage scenarios for IPv6 deployment over MPLS:

■ Where an MPLS network exists, the method may be most appropriate.
■ If hardware upgrade for line rate IPv6 is expensive, running IPv6-over-IPv4/MPLS may be a good compromise for line rate IPv6 (avoiding encapsulation in software tunnels).

Backbone networks that have already deployed MPLS might consider several IPv4-IPv6 migration strategies:

■ Native IPv6 over MPLS: In this scenario, IPv6 transport over an MPLS network is completely symmetric to the IPv4 case. It requires that all routers in the MPLS network become dual stack and use IPv6 routing protocols (both interior and exterior) together with IPv6-enabled Label Distribution Protocol (LDP); however, major router vendors seem to have no plans for adding IPv6 support to LDP in a foreseeable future.
■ L2 tunneling over MPLS [MAR200601]. Entire L2 frames (e.g., Ethernet with Institute of Electrical and Electronics Engineers (IEEE) 802.1q encapsulation, ATM Adaptation Layer 5 (AAL5), etc.) are switched across the MPLS core; hence, the L3 protocol is completely transparent. This feature is available, in one form or another, on most major routing platforms, including Cisco IOS and Juniper JunOS; however, L2 tunneling over MPLS brings nothing new from the perspective of IPv6 and, moreover, is not really suitable for wide-area networks.

■ IPv6 over IPv4/MPLS core: This method relies on the distribution of IPv6 prefixes (and corresponding labels) among the edge label switch routers (LSRs) using standard Border Gateway Protocol version 4 (BGPv4) over IPv4, where the next hop is identified by an IPv4 address. Cisco Systems implements this functionality under the name 6PE (IPv6 provider edge router).

8.2.5 Layer 2 Transport Protocol Considerations for NRENs

This section addresses Layer 2 considerations in NREN when transitioning to IPv6. Layer 2 technologies considered are Packet over SONET (PoS), MPLS, and ATM.

8.2.5.1 Packet over SONET (PoS) Scenario

In the case of an IPv4 PoS network, NREN envisions two different approaches:

■ Dual-stack operation on the NREN core routers as a natural path forward, given appropriate robust code and appropriate studies into the management and operational implications.
■ Deployment of a parallel, nondisruptive IPv6 network.

However, NREN does not expect parallel infrastructures to be long-term transition solutions, and the dual-stack, single infrastructure method is preferred. Ideally, in dual stack, both IPv4 and IPv6 are handled natively at line rate, but this may not be the case. It depends on the vendor/platform architecture; however, as new vendor hardware is released over time, the performance has improved such that, for the majority of networks, those with up-to-date hardware, the performance of IPv6 and IPv4 is equivalent.

Another consideration is how long term the dual-stack mode of operation would be and what would be its exit strategy (e.g., run IPv6 with IPv4 carried as tunneled traffic in the IPv6-only backbone). That scenario is, however, some time away (probably the next decade).

8.2.5.2 MPLS Scenario

An NREN may have an existing MPLS network and there are some specific options for running IPv6 over MPLS ([RFC4798] for example). MPLS would not normally be deployed purely to enable IPv6; however if the upgrade path for hardware to enable IPv6 at line rate is expensive, MPLS is an option to carry IPv6 natively at line rate.

8.2.5.3 ATM Scenario

The NREN could also have an ATM network and thus be able to run IPv6 over parallel PVCs. However, no NRENs are currently using this method, and ATM would not be deployed purely to enable IPv6.

8.3 IPv6 Operations Working Group: Transition Scenarios for ISPs

As mentioned, this summary is based on [RFC4029]. Overall, the RFC contains many similar issues to the NREN document. (Note that members of 6NET have contributed to the RFC.) The RFC begins by describing stages of transition that an ISP network may go through during transition and notes that an ISP may transition different elements of its network at different times (e.g., it may start by offering IPv6 in IPv4 tunnels to early-adopter customers ahead of a full dual-stack service). The RFC identifies four stages:

- 1: Launch: The ISP obtains an IPv6 SubTLA address.
- 2a: Backbone: The ISP upgrades its backbone to carry IPv6 traffic.
- 2b: Customer connection: The ISP customer(s) connection is upgraded to carry IPv6 traffic.
- 3: Complete: IPv6 is available through the ISP and customer network.

The RFC suggests that the ISP backbone upgrade can happen as new procurements are made. Clearly, a recommendation to procure IPv6-capable equipment is wise, but some platforms have variations in line card capabilities that may be subtle to the buyer. For example, some NRENs in 6NET have noted that having dual stack on certain hardware only recently bought can be a problem, and that additional expensive upgrades may be required sooner than expected. As a result, an ISP may choose to use MPLS on existing hardware to carry IPv6 at line rate rather than doing a "double upgrade."

The RFC considers different backbone technologies:

- IPv4: In this case, tunnels may be used initially, to be replaced with dual-stack networking as it becomes available.
- MPLS: This can be deployed as native IPv6-over-MPLS or as the IPv6-over-IPv4/MPLS method described in [CLE200601]. Ideally MPLS networks should deploy native IPv6 routing and forwarding or use IPv6 Label Switched Paths (LSPs), but using tunneling over IPv4 LSPs or through the [CLE200601] approach (also known as 6PE) offers an interim step.
- ATM is not discussed in the RFC. Given its general absence from NREN networks now, this seems appropriate for most cases with ISPs.

The RFC considers Open Shortest Path First version 2 (OSPFv2) and intermediate system to intermediate system (IS-IS) as the possible IPv4 Interior Gateway Protocols (IGPs) before transition. Routing Information Protocol version 2 (RIPv2) and Internal Border Gateway Protocol (iBGP) are not considered beyond point-to-point routing. The choice presented in the RFC is the same as the NRENs have generally faced, however: OSPFv3 or IS-IS for IPv6. The basic issue then is whether to have separate routing processes for each protocol, IPv4 and IPv6.

Separate processes have more overhead but offer separation and thus resilience between IPv4 and IPv6 stability. The possible combinations are:

- OSPFv2 for IPv4, IS-IS for IPv6
- OSPFv2 for IPv4, OSPFv3 for IPv6
- IS-IS for IPv4, OSPFv3 for IPv6
- IS-IS for both IPv4 and IPv6 (same process)
- IS-IS with separate processes (databases), one instance for each protocol

The RFC recommends IS-IS for IPv4 and IPv6, a method already used by three of the NRENs. The deployment and usage of these protocols is also discussed for NRENs in [6NET-D312].

Multicast is somewhat dismissed by the RFC but has formed a major part of the work of 6NET. For discussion on IPv6 multicast routing, see [6NET-D312].

The RFC then discusses customer access. For small customer sites, the connectivity mechanisms are categorized as "managed" or "opportunistic." The former consist of native service or a configured tunnel; the latter include 6to4 and, for example, Teredo — they provide "short-cuts" between nodes using the same mechanisms and are available without contracts with the ISP. The ISP may offer opportunistic services, mainly a 6to4 relay, especially as a test when no actual service is offered yet. At the later phases, ISPs might also deploy 6to4 relays and Teredo servers (or similar) to optimize their customers' connectivity to 6to4 and Teredo nodes.

The RFC does include consideration of large end sites, which usually have a managed network. Dual-stack access service is a possibility as the customer network is managed (although customer premises equipment (CPE) upgrades may be necessary). Configured tunnels are a good solution when a network address translation (NAT) is not in the way and the IPv4 end-point addresses are static. In this scenario, NAT traversal is not typically required. Teredo is not applicable in this scenario as it can only provide IPv6 connectivity to a single host, not the whole site. 6to4 is not recommended due to its reliance on the relays and provider-independent address space, which makes it impossible to guarantee the required service quality and manageability large sites typically want.

The RFC cites three example scenarios for transition analysis.

- The first is an xDSL (Digital Subscriber Line) provider with a recommendation e.g., to deploy a 6to4 relay for optimized customer 6to4 performance.
- The second example is an IPv4 MPLS network, similar to the discussion in Section 8.2.4.
- The third example is of a transit provider offering IP connectivity to other providers but not to end users or enterprises. In this case, the whole backbone can be upgraded to dual stack in a reasonably short time after a trial with a couple of routers. The provider can also offer, at a first phase only, a configured tunnel service with BGP peering to its significant sites and customers.

Finally, the RFC closes with security considerations in three areas:

- Generic best practices
- Security concerns through complexity of transition tools (e.g., open 6to4 relays)
- Complexity in managing a dual stack (e.g. access control lists (ACLs)) and in the ability to trace customers (e.g., where stateless address autoconfiguration is used or, perhaps more importantly, RFC 3041 addresses)

References

[CLE200601] J. De Clercq, D. Ooms, S. Prevost, and F. Le Faucheur, Connecting IPv6 Islands over IPv4 MPLS Using IPv6 Provider Edge Routers (6PE), draft-ooms-v6ops-bgp-tunnel-07.txt, December 12, 2006.

[MAR200601] L. Martini, E. C. Rosen, and N. El-Aawar, Encapsulation Methods for Transport of Layer 2 Frames over MPLS Networks, draft-martini-l2circuit-encap-mpls-12.txt, September 2006.

[6NE200501] 6NET, D2.2.4: Final IPv4 to IPv6 Transition Cookbook for Organisational/ISP (NREN) and Backbone Networks, version 1.0, February 4, 2005, Project Number IST-2001-32603, CEC Deliverable Number 32603/UOS/DS/2.2.4/A1.

[6NET-D312] IPv6 Cookbook for Routing, DNS, Intra-domain Multicast, Inter-domain Multicast and Security, 2nd version, 6NET Project Deliverable D3.1.2.

[6NET-D622] Operational Procedures for Secured Management with Transition Mechanisms, 6NET Project Deliverable D6.2.2.

[RFC4029] M. Lind, V. Ksinant, S. Park, A. Baudot, and P. Savola, Scenarios and Analysis for Introducing IPv6 into ISP Networks, RFC 4029, March 2005.

[RFC4798] J. De Clercq, D. Ooms, S. Prevost, F. Le Faucheur, Connecting IPv6 Islands over IPv4 MPLS Using IPv6 Provider Edge Routers (6PE), RFC 4798, February 2007.

Chapter 9

IPv6 Applications

9.1 Introduction

This chapter discusses some basic application-related issues that have to be taken into consideration when migrating existing business applications to an Internet Protocol version 6 (IPv6) networking environment.

9.2 Application Programming Interface Overview

Applications may use either Transmission Control Protocol (TCP) or User Datagram Protocol (UDP) for communicating with remote hosts on the Internet. Both are services that work on top of the IP network protocol. TCP is a reliable "streams" service that requires a connection establishment phase when a host is making a connection to a remote server host. UDP is an unreliable datagram service and does not require any connection establishment before sending data.

The services provided by TCP and UDP to applications are in general referred to in terms of primitives and their associated parameters, with the parameters including items such as the source and destination ports and IP addresses. Figure 9.1 illustrates this relationship for the IPv4 protocol stack, showing the primitives and parameters for some of the interactions between the application and the TCP and IP layers. The example shown is for an application opening a TCP connection to a remote IP address and TCP port with the Open primitive and subsequently receiving an indication that the connection was opened (via the Open ID primitive). Subsequent primitives would be issued for streaming the TCP segments over the

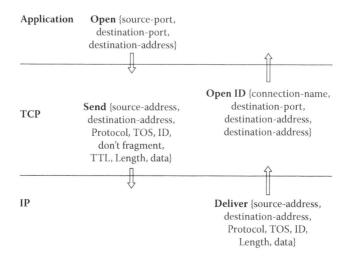

Figure 9.1   Sample TCP primitives and parameters.

existing connection via TCP-related Send and Deliver primitives (not shown on the figure). The IP layer primitives shown in the figure are concerned only with sending and receiving particular datagrams and use the IP-related Send and Deliver primitives shown in the figure. For IPv6, the basic TCP and IP primitives — to open and close a connection and to send and receive data — will remain the same, but some new parameters, such as the Flow Label, need to be added. In addition, the structure of some of the parameters will need to change. For example, the IP address field will have to change from the 32-bit IPv4 address to the 128-bit IPv6 address. Thus, from an application programming point of view, the only differences between network protocols are the address schemes used and some of the associated parameters. For TCP/IP, a key requirement of the application programming interface (API) is that it understand both IPv4 and IPv6 addresses. Otherwise, abstract operations such as connect, send, deliver, and disconnect will then hide changes from the application. In this way, the use of the abstraction of primitives and parameters can minimize the changes needed to applications when moving from IPv4 to IPv6.

The actual form of a primitive is implementation dependent and will vary with the particular operating system of the end system. Examples of the implementation of primitives are subroutines and function calls. An example of the sockets implementation, used in a number of operating systems, including UNIX, is presented in Section 9.3. The section addresses the changes required to make UNIX applications independent of the underlying IP layer.

The UDP protocol is at the same layer as TCP but provides a much simpler connectionless service for application procedures. The associated primitives for UDP are thus much simpler, adding only a port-addressing capability to the underlying IP layer services. The UDP primitives are illustrated in Figure 9.2.

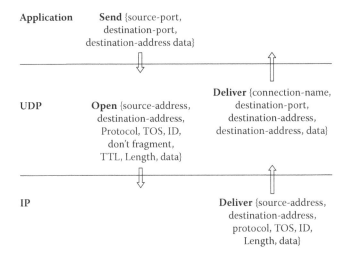

Figure 9.2   Sample UDP primitives and parameters.

Application Layer (user level)	
Socket Layer	
Stream Sockets	Datagram Sockets
TCP	UDP
Internet Protocol (IP)	
Network Interface	

Figure 9.3   Socket architecture.

9.3 Socket API Example

A *socket* is a loose term used to describe an endpoint for communication and is the de facto standard API for TCP/IP and UDP/IP. The socket layer is shown in Figure 9.3, in which the sockets provide a stream socket service, based on TCP, and a datagram socket service, based on UDP.

[RFC3493] describes extensions needed to the socket interface to include IPv6 support in addition to IPv4. Most of the basic communication functions need not change for IPv6, but required extensions include a new socket address structure to carry IPv6 addresses, new address conversion functions, and some new socket options. These extensions are designed to provide access to the basic IPv6 features required by TCP and UDP applications while introducing a minimum of change into the system and providing complete compatibility for existing IPv4 applications. Thus, with these extensions, one can code for IPv6 with the extended API, and the applications will still work when IPv6 is enabled.

The extensions described in [RFC3493] for IPv6 cover four areas:

- Core socket functions
- Address data structures
- Name-to-address translation functions
- Address conversion functions

9.3.1 Core Socket Functions

The original core socket functions, those that set up and tear down connections and send and receive packets, were designed to be protocol independent. The new API, like the old, remains protocol independent, so it does not matter if the network protocol is IPv4 or IPv6, the same code will work with no changes. These core functions do not have to change for IPv6, but they do require some new IPv6 data structures, which are described in the next section.

9.3.2 Address Data Structures

The key is that protocol addresses passed as function arguments are passed via pointers. The pointer relates to a protocol-specific address data structure defined for each protocol that the socket functions support. Applications must cast pointers to these protocol-specific address structures into pointers to the generic "sockaddr" address structure when using the socket functions. The generic socket address structure is defined as follows:

```
struct sockaddr {
  unsigned short sa_family;  /*Address family (e.g., AF_INET)*/
  char sa_data[14];          /*Protocol-specific address information*/
};
```

Here, the *sa_family* field specifies the type of protocol with, for example, AF_INET indicating IPv4. The remaining 14 bytes of this structure are always protocol dependent.

Although socket functions handle the generic socket address structure, application developers must include the socket address structure according to the communication protocol used to establish the socket. A protocol-specific data structure is defined for each protocol that the socket functions support. When calling the core socket functions, the protocol-specific data structure for IPv4 is cast to the *sockaddr_in address* structure, which includes IPv4 addresses and port numbers.

Generic Sockaddr	Family	Protocol Dependent Address Information		
	2 bytes	2 bytes	4 bytes	8 bytes
Sockaddr_in	AF_INET	Port	IPv4 Address	Not used

Figure 9.4   Generic socket and IPv4 socket address structures.

```
struct sockaddr_in{
short sin_family;          /* AF_INET for IPv4 address family*/
unsigned short sin_port;   /* Port number (16 bits). */
struct in_addr sin_addr;   /* Internet address. (32 bits) */
char sin_zero[8];
};
```

Here, the *sin_family* field is an indicator of the address family, which in this case is AF_INET for IPv4. The *sin_port* field is a 16-bit unsigned value used to represent a port number, and the *in_addr* struct is exactly four bytes long to accommodate an IPv4 address. Finally, eight bytes are unused. The relationship of the generic *sockaddr* structure and the IPv4 *sockaddr_in* structure is shown in Figure 9.4.

The *sockaddr_in* structure is the protocol-specific data structure for IPv4; however, this data structure is not large enough to hold the 16-octet IPv6 address as well as the other information (address family and port number) that is needed. So, a new address data structure must be defined for IPv6. The protocol-specific data structure for IPv6 is *sockaddr_in6*.

```
struct sockaddr_in6 {
    sa_family_t sin6_family;   /* AF_INET6 for IPv6 address family*/
    in_port_t sin6_port;       /* Transport layer port # */
    uint32_t sin6_flowinfo;    /* IPv6 flow information */
    struct in6_addr sin6_addr; /* IPv6 address */
    uint32_t sin6_scope_id;    /* IPv6 scope-id */
};
```

Existing applications written assuming IPv4 using the *sockaddr_in* structure can be easily ported by changing this structure to *sockaddr_in6*. Notice that the *sockaddr_in6* structure will normally be larger than the generic *sockaddr* or *sockaddr_in* structures. Any existing code that makes the assumption that these structures are of equal length needs to be examined carefully when converting to IPv6.

The API also specifies a number of compatibility features for transitioning to IPv6. For example, the ability for IPv6 applications to interoperate with IPv4 applications using the IPv4-mapped IPv6 address format is addressed by the RFC. With this format, the IPv4 address is encoded into the low-order 32 bits of the IPv6 address, written as follows:

```
::FFFF:<IPv4-address>
```

Applications may use AF_INET6 sockets by simply encoding the destination's IPv4 address as an IPv4-mapped IPv6 address and passing that address, within a *sockaddr_in6* structure, in the TCP connect() or UDP sendto() call.

9.3.3 Name-to-Address Translation Functions

The most common function for translating names to addresses in IPv4 is gethost-byname(). It is still retained for backward compatibility but is inadequate for IPv6 because there is no way to specify anything about the types of addresses desired (IPv4, IPv6, and so on).

The getaddrinfo() function, defined in [RFC3493], solves these problems. The getaddrinfo function returns a linked list of addrinfo structures that contains information requested for a specific set of hostname, service, and additional information. Applications can examine the linked list returned by getaddrinfo() and use the appropriate structure.

9.3.4 Address Conversion Functions

Typically, the functions inet_addr() and inet_ntoa() are used to convert an IPv4 address between binary and text form. The analogous IPv6 functions are called inet_pton() and inet_ntop(), respectively. The inet_pton() function converts a text address to its binary equivalent; inet_ntop() does the reverse, converting a binary address to printable text. Because both of these functions have a parameter for specifying the address family, they can be used to convert both IPv4 and IPv6 addresses.

9.3.5 Socket Functions for IPv6

Applications call the socket() function to create a socket descriptor that represents a communication endpoint. The basic socket call is

```
s = socket (family, type, protocol);
```

The arguments to the socket() function tell the system which protocol to use and which format address structure will be used in subsequent functions. For example, to create an IPv4/TCP socket, applications make the call:

```
s = socket(AF_INET, SOCK_STREAM, 0);
```

with the protocol field set to 0 indicating the IP protocol.

To create an IPv4/UDP socket, applications make the call:

```
s = socket(AF_INET, SOCK_DGRAM, 0);
```

Applications may create IPv6/TCP and IPv6/UDP sockets by simply using the constant AF_INET6 instead of AF_INET in the first argument. For example, to create an IPv6/TCP socket, applications make the call:

```
s = socket(AF_INET6, SOCK_STREAM, 0);
```

To create an IPv6/UDP socket, applications make the call:

```
s = socket(AF_INET6, SOCK_DGRAM, 0);
```

Once the application has created an AF_INET6 socket, it must use the *sockaddr_in6* address structure when passing addresses in to the system.

9.4 IPv6 Support for Networking Applications

A number of network-related application or utility programs have been developed in support of IPv4. Much work has been completed to port these applications to IPv6. One example of this work is in [BIE200601], in which the status of these applications is tracked for the LINUX operating system. The number of supported applications is significant, with most important networking applications providing native IPv6 support. The reference contains the status and source for the following network services:

Port	Service		Port	Service
7/tcp	echo		53/udp	domain
9/tcp	discard		69/udp	tftp
13/tcp	daytime		79/tcp	finger
19/tcp	chargen		80/tcp	http
21/tcp	ftp		109/tcp	pop2
22/tcp	ssh		110/tcp	pop3
23/tcp	telnet		113/tcp	auth
25/tcp	smtp		119/tcp	nntp
43/tcp	whois		123/udp	ntp

Port	Service
143/tcp	imap4
194/tcp	irc
220/tcp	imap3
389/tcp	ldap
443/tcp	https
512/tcp	exec
513/tcp	login
514/tcp	cmd

Port	Service
631/tcp	ipp
636/tcp	ldaps
873/tcp	rsync
993/tcp	imaps
995/tcp	pop3s
5222/tcp	jabber
6667/tcp	ircd

References

[BIE200601] P. Bieringer, F. Baraldi, S. Piunno, M. Tortonesi, E. Toselli, and D. Tumiati, Current Status of Ipv6 Support for Network Applications, http://www.deepspace6.net/docs/ipv6_status_page_apps.html.

[RFC3493] R. Gilligan, S. Thomson, J. Bound, J. McCann, and W. Stevens, Basic Socket Interface Extensions for IPv6, RFC 3493, February 2003.

Chapter 10

Security in IPv6 Networks

10.1 Introduction

This chapter identifies security mechanisms that can be employed in Internet Protocol version 6 (IPv6) environments. Many of these mechanisms are imported from the IPv4 world. It is not the goal of this chapter to provide a tutorial on security but simply to identify some of the commonly available techniques, methods, and protocols that can be used to secure IPv6 networks. For the purpose of this discussion, we view security in four realms: (1) confidentiality and integrity of information while in transit (in a network); (2) perimeter/access security in reference to a defined set of private information technology (IT) assets; (3) system security/integrity (including both client and server systems); and (4) security of data at rest (namely, security of database/storage systems).[1] However, this chapter only deals with the first item. This kind of security can be seen as achievable using tunnels that carry encrypted information.

10.2 Confidentiality and Integrity of Information While in Transit

To illustrate the gamut of possibilities related to confidentiality and integrity of information while in transit, Figure 10.1 depicts the positioning of the encryption

---

[1] More generally, security practitioners see security as covering confidentiality (which deals with measures to provide encryption, authentication [including digital certificates], and authorization), integrity (which deals with measures to prevent corruption and support validation via digital signatures), and availability (which deals with measures to avoid denial of service).

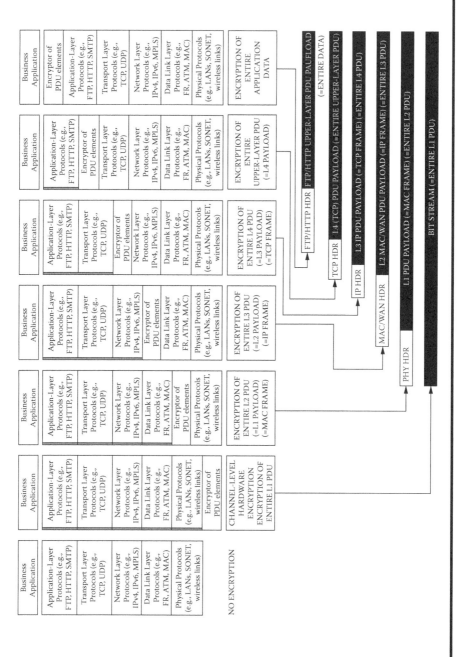

Figure 10.1 Encryption tunneling modes.

function within the protocol stack of a communicating node. Although military applications often use link-level encryption at the physical layer (channel), commercial applications tend to use encryption (tunneling) at the network layer (via IPsec tunneling, discussed below) or at the transport layer (via Transport Layer Security (TLS)/Secure Sockets Layer (SSL) tunneling, also discussed below).

As noted, the confidentiality and integrity of information while in transit (in an IPv6 network) is achieved by using encryption (tunneling) at the network layer via IPsec tunneling or at the transport layer via TLS/SSL tunneling. The tunnel is generally established between end-system nodes (e.g., between a client accessing a remote host) or between network-edge nodes (e.g., wide area network router in the intranet to wide area network router at the far end of the intranet). In some cases, the telecommunications service provider (rather than the end-user organization) supports the edge-to-edge encryption function by deploying premises-based provider equipment (PE) that they manage on behalf of the user. The resulting encrypted/tunnel arrangement is called a virtual private network (VPN), particularly when the service provider handles the encryption or when the Internet is used (and the end-user organization does the router-to-router encryption). VPN is a generic term that covers the use of public or private networks to create groups of users separated from other network users and may communicate among them as if they were on a private network [RFC2764]. It should be noted in passing that VPNs can also be realized at Layer 2. Figure 10.2 provides a classification of VPNs as outlined in [RFC4026].

Encryption is the cryptographic transformation of data (called *plaintext*) into a form (called *ciphertext*) that conceals the data's original meaning to prevent it from being known or used. It is the cryptographic transformation of data to produce

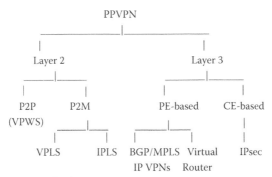

PE = Provider Equipment
CE = Customer Equipment
PPVPN = Provider-Provisioned Virtual Private Network

Figure 10.2   Classification of VPNs.

ciphertext [ISO198401]. If the transformation is reversible, the corresponding reversal process is called *decryption*, which is a transformation that restores encrypted data to its original state. Usually, the plaintext input to an encryption operation is cleartext (in some cases, the plaintext may be ciphertext that was output from another encryption operation). Encryption and decryption involve a mathematical algorithm for transforming data. In addition to the data to be transformed, the algorithm has one or more inputs that are control parameters: a key value that varies the transformation and, in some cases, an initialization value that establishes the starting state of the algorithm [RFC2828].

10.3 IPsec Mechanisms

IPsec as defined in [RFC2401] is a fundamental element of IPv6 security in the context of confidentiality and integrity of information while in transit. IPsec is a protocol that provides for the support of encrypted payloads in IP networks; it includes encryption and authentication technologies. It is a broadly deployed mechanism (in IPv4 environments) used to support VPNs defined over the Internet.

The IPsec architecture specifies (1) security protocols (Authentication Header (AH) and Encapsulating Security Payload (ESP)); (2) security associations (what they are, how they work, how they are managed, and associated processing); (3) key management (IPsec key exchange (IKE)); and (4) algorithms for authentication and encryption. The set of security services includes access control service, connectionless data integrity service, data origin authentication service, protection against replays (detection of the arrival of duplicate datagrams, within a constrained window), data confidentiality service, and limited traffic flow confidentiality [RFC2828].

AH is designed to provide connectionless data integrity service and data origin authentication service for IP datagrams and (optionally) to provide protection against replay attacks. AH provides for integrity but without confidentiality. AH may be used alone, in combination with the IPsec ESP protocol, or in a nested fashion with tunneling. Security services can be provided between a pair of communicating hosts, between a pair of communicating security gateways, or between a host and a gateway.

ESP can provide the same security services as AH, and ESP can also provide data confidentiality service. ESP provides for confidentiality with optional integrity and authentication. The main difference between authentication services provided by ESP and AH is the extent of the coverage; ESP does not protect IP header fields unless they are encapsulated by AH [RFC2828].

The granularity of security protection in IPsec is at the datagram level. IPsec treats everything in an IP datagram after the IP header as one integrity unit. Usually, an IP datagram has three consecutive parts: the IP header (for routing purpose only), the upper-layer protocol headers (e.g., the Transmission Control Protocol

(TCP) header), and the user data (e.g., TCP data). In transport mode,[2] an IPsec header (AH or ESP) is inserted after the IP header and before the upper-layer protocol header to protect the upper-layer protocols and user data. In tunnel mode, the entire IP datagram is encapsulated in a new IPsec packet (a new IP header followed by an AH or ESP header). In either mode, the upper-layer protocol headers and data in an IP datagram are protected as one indivisible unit. The keys used in IPsec encryption and authentication are shared only by the sender-side and receiver-side security gateways. All other nodes in the public Internet, whether they are legitimate routers or malicious eavesdroppers, see only the IP header and will not be able to decrypt the content or tamper with it without detection. Traditionally, the intermediate routers do only one thing — forward packets based on the IP header (mainly the destination address field); IPsec's "end-to-end" protection model is well suited to this layering paradigm [ZHA200401].

As implied above, a VPN is a restricted-use, logical (i.e., artificial or simulated) computer network that is constructed from the system resources of a relatively public, physical (i.e., real) network (such as the Internet), often by using encryption (located at hosts or gateways), and often by tunneling links of the virtual network across the real network [RFC2764]. For example, if a corporation has local area networks (LANs) at several different sites, each connected to the Internet by a firewall, the corporation could create a VPN by (1) using encrypted tunnels to connect from firewall to firewall across the Internet and (2) not allowing any other traffic through the firewalls. A VPN is generally less expensive to build and operate than a dedicated real network, because the virtual network shares the cost of system resources with other users of the real network [RFC2828]. Having noted that IPsec is fairly important in the context of network security (at least in terms of confidentiality and integrity), we identify the following list of relevant RFCs[3] for IPsec:

- RFC 2401: Security Architecture for the Internet Protocol
- RFC 2402: IP Authentication Header (AH)
- RFC 2403: The Use of HMAC-MD5-96 within Encapsulating Security Payload (ESP) and Authentication Header (AH)
- RFC 2404: The Use of HMAC-SHA-1-96 within Encapsulating Security Payload (ESP) and Authentication Header (AH)
- RFC 2405: The ESP DES-CBC (Cipher Block Chaining) Cipher Algorithm with Explicit IV (Initialization Vector)
- RFC 2406: IP Encapsulating Security Payload (ESP)
- RFC 2407: The Internet IP Security Domain of Interpretation for ISAKMP

---

[2] Transport mode and tunnel mode are described in the paragraphs that follow.
[3] RFCs describe "quasi specifications" (which may in due course progress to full standards) for networking (and occasionally computing) approaches that may be used by vendors to develop interoperable products.

■ RFC 2408: Internet Security Association and Key Management Protocol (ISAKMP)
■ RFC 2409: The Internet Key Exchange (IKE)
■ RFC 2410: The NULL Encryption Algorithm and Its Use with IPsec
■ RFC 2411: IP Security Document Roadmap
■ RFC 2412: The OAKLEY Key Determination Protocol
■ RFC 3602: The AES-CBC Cipher Algorithm and Its Use with IPsec
■ RFC 3776: Using IPsec to Protect Mobile IPv6 Signaling Between Mobile Nodes and Home Agents

Next we look at some of the key capabilities required for secure communications in IPv6 networks.

10.3.1 Keyed Hashing for Message Authentication

The technical details described in specification RFC 2104, as they refer to hashing, must generally be supported in a network environment (intranet, extranet, or institutional) that makes use of untrusted transmission links (this is the case because RFC 2104 is referenced by RFC 2403, which describes how IPsec protects the integrity of datagrams). A keyed hash that can be based on any iterated cryptographic hash (e.g., message digest 5 [MD5] or secure hash algorithm 1 (SHA-1)). The cryptographic strength of the hashed message authentication code (HMAC) depends on the properties of the selected cryptographic hash. The goals of HMAC are as follows [RFC2828]:

1. To use available cryptographic hash functions without modification, particularly functions that perform well in software and for which software is freely and widely available
2. To preserve the original performance of the selected hash without significant degradation
3. To use and handle keys in a simple way
4. To have a well-understood cryptographic analysis of the strength of the mechanism based on reasonable assumptions about the underlying hash function
5. To enable easy replacement of the hash function in case a faster or stronger hash is found or required

10.3.2 Security Architecture for the Internet Protocol

The technical details described in specification [RFC2104], as they refer to transport mode and tunnel mode operation, must generally be supported.

Transport mode is an IPsec mode as defined in [RFC2401], Security Architecture for the Internet Protocol. Transport mode is allowed between two end hosts only; tunnel mode is required when at least one of the endpoints is a security gateway (intermediate system that implements IPsec functionality, e.g., a router) [RFC3884]. Transport mode secures portions of the existing IP header and the payload data of the packet and inserts an IPsec header between the IP header and the payload. In transport mode, IPsec inserts a security protocol header into outgoing IP packets between the original IP header and the packet payload. The contents of the IPsec header are based on the result of a security association (SA) lookup that uses the contents of the original packet header as well as its payload (especially transport layer headers) to locate an SA in the security association database (SAD). When receiving packets secured with IPsec transport mode, a similar SA lookup occurs based on the IP and IPsec headers, followed by a verification step after IPsec processing that checks the contents of the packet and its payload against the respective SA. The verification step is similar to firewall processing [RFC3884].

Tunnel mode is an IPsec mode as defined in [RFC2401], Security Architecture for the Internet Protocol. Tunnel Mode is required when at least one of the endpoints is a security gateway (intermediate system that implements IPsec functionality, e.g., a router). By contrast, transport mode is allowed between two end hosts only [RFC3884]. Transport mode secures portions of the existing IP header and the payload data of the packet and inserts an IPsec header between the IP header and the payload; tunnel mode adds an additional IP header before performing similar operations. When using tunnel mode, IPsec prepends an IPsec header and an additional IP header to the outgoing IP packet. In essence, the original packet becomes the payload of another IP packet, which IPsec then secures. This has been described as "a tunnel mode SA is essentially a (transport mode) SA applied to an IP tunnel." In IPsec tunnel mode, the IP header of the original outbound packet together with its payload (especially transport headers) determine the IPsec SA, as for transport mode. However, a tunnel mode SA also contains encapsulation information, including the source and destination IP addresses for the outer tunnel IP header, which is also based on the original outbound packet header and its payload [RFC3884].

10.3.2.1 IP Authentication Header

For the IP AH, the technical details described in specification [RFC2402] for the support of AH must generally be supported.

10.3.2.2 Use of HMAC-MD5-96 within ESP and AH

For use of HMAC-MD5-96 within ESP and AH, the technical details described in specification [RFC2403] in reference to hashing must generally be supported.

10.3.2.3 Use of HMAC-SHA-96 within ESP and AH

Regarding use of HMAC-SHA-96 within ESP and AH, the technical details described in specification RFC 2404 in reference to hashing must generally be supported.

10.3.2.4 ESP DES-CBC Cipher Algorithm with Explicit IV

For ESP Data Encryption Standard Cipher Block Chaining (DES-CBC) cipher algorithm with Explicit IV, the technical details described in specification RFC 2405 must generally be supported; however, stronger algorithms than DES should be used, such as the Advanced Encryption Standard (AES).

DES is a U.S. government standard (developed at IBM in 1977) that specifies the encryption algorithm and the policy for using the algorithm to protect unclassified, sensitive data. DES is specified in the American National Standards Institute (ANSI) X3.92 and X3.106 standards and in the FIPS (Federal Information Processing Standard) 46 and 81 standards. The algorithm was adopted by the National Institute of Standards and Technology (NIST) and has been widely used for public and government applications. Originally, the algorithm was judged so difficult to break by the U.S. government that it was restricted for exportation to other countries. DES is a block cipher with a 56-bit key and an eight-byte block size. DES applies a 56-bit key to each 64-bit block of data. DES-encrypted messages are now vulnerable to deciphering considering the advancements in computing, particularly grid computing.

Many organizations now use Triple DES. The Triple DES algorithm is a variation proposed for ESP that uses a 168-bit key, consisting of three independent 56-bit quantities used by DES and a 64-bit initialization value. Triple DES is a block cipher, based on DES, that transforms each 64-bit plaintext block by applying the DES three successive times, using either two or three different keys, for an effective key length of 112 or 168 bits. Each datagram contains an initialization vector (IV) to ensure that each received datagram can be decrypted even when other datagrams are dropped or a sequence of datagrams is reordered in transit.

AES is a specific encryption standard (also known as Rijndael) that is a symmetric block cipher adopted by the U.S. government. The standard was adopted by NIST as U.S. FIPS PUB 197 in 2001 after a five-year standardization process. The cipher was developed by two Belgian cryptographers, Vincent Rijmen and Joan Daemen, and submitted to the AES selection process under the name Rijndael. The block size is 128 bits; the key typically is 128, 192, or 256 bits. AES is intended as a more robust replacement for the DES and for the Triple DES. The algorithm is royalty free and offers security of a sufficient level to protect data for the next 20 to 30 years. In addition to U.S. FIPS PUB 197, key AES documents are as follows:

■ RFC 3268: Advanced Encryption Standard Ciphersuites for Transport Layer Security (TLS)

- RFC 3394: Advanced Encryption Standard Key Wrap Algorithm
- RFC 3537: Wrapping a Hashed Message Authentication Code (HMAC) key with a Triple-Data Encryption Standard (DES) Key or an Advanced Encryption Standard (AES) Key
- RFC 3565: Use of the Advanced Encryption Standard Encryption Algorithm in Cryptographic Message Syntax (CMS)
- RFC 3566: The AES-XCBC-MAC-96 Algorithm and Its Use with IPsec
- RFC 3602: The AES-CBC Cipher Algorithm and Its Use with IPsec
- RFC 3664: The AES-XCBC-PRF-128 Algorithm for the Internet Key Exchange Protocol (IKE)
- RFC 3686: Using Advanced Encryption Standard Counter Mode with IPsec Encapsulating Security Payload (ESP)
- RFC 3826: The Advanced Encryption Standard Cipher Algorithm in the SNMP User-Based Security Model
- RFC 3962: Advanced Encryption Standard Encryption for Kerberos 5

10.3.2.5 IP Encapsulating Security Payload

For IP ESP, the technical details described in specification [RFC2406] related to encapsulation must generally be supported.

10.3.2.6 Automatic Key Management

Automatic key management, as described in [RFC2407], [RFC2408], and [RFC2409], is not a mandatory requirement of the IP Security Architecture. Note, however, that in some environments (e.g., cellular/3G environments) the IP addresses of a host may change dynamically, and the use of manually configured security associations is not practical. Some applications could use the IKE mechanism for key management; other applications could use different mechanisms.

10.3.2.7 The Internet Key Exchange

IKE described in [RFC2409] is optional according to the IPv6 specifications, as noted above, but may be necessary in some applications.

Note that interactions with the Internet Control Message Protocol version 6 (ICMPv6) packets and IPsec policies may cause unexpected behavior for IKE-based SA negotiation unless some special handling is performed in the implementations. The ICMPv6 provides many functions, which in IPv4 were either nonexistent or provided by lower layers. For instance, IPv6 implements address resolution using an IP packet, ICMPv6 neighbor solicitation message. In contrast, IPv4 uses an Address Resolution Protocol (ARP) message at a lower layer [RFC3316].

The IPsec architecture has a security policy database that specifies which traffic is protected and how. It turns out that the specification of policies in the presence of ICMPv6 traffic is not easy. For instance, a simple policy of protecting all traffic between two hosts on the same network would trap even address resolution messages, leading to a situation for which IKE cannot establish an SA because to send the IKE UDP packets one would have had to send the neighbor solicitation message, which would have required an SA. To avoid this problem, neighbor solicitation, neighbor advertisement, router solicitation, and router advertisement messages must not lead to the use of IKE-based SA negotiation. The redirect message should not lead to the use of IKE-based SA negotiation. Other ICMPv6 messages may use IKE-based SA negotiation as is desired in the security policy database. All of this limits the usefulness of IPsec in protecting all ICMPv6 communications [RFC3316].

10.4 Transport Layer Security Mechanisms

In general, the need to use a security mechanism depends on the intended application for it. Different security mechanisms are useful in different contexts and have different limitations. Some applications require the use of TLS [RFC2246]; in some situations, IPsec [RFC2401] is used [RFC3316].

The primary goal of TLS is to provide privacy and data integrity between two communicating applications. TLS is the successor protocol to SSL, developed by IETF. It can be used for general communication authentication and encryption (TLS version 1 is similar to SSL version 3). The protocol is composed of two layers: the TLS Record Protocol and the TLS Handshake Protocol. At the lowest level, layered on top of some reliable transport protocol (e.g., TCP), is the TLS Record Protocol. The TLS Record Protocol provides connection security that has two basic properties [RFC2246]:

■ The connection is private. Symmetric cryptography is used for data encryption (e.g., DES). The keys for this symmetric encryption are generated uniquely for each connection and are based on a secret negotiated by another protocol (such as the TLS Handshake Protocol). The Record Protocol can also be used without encryption.

■ The connection is reliable. Message transport includes a message integrity check using a keyed Message Authentication Code (MAC). Secure hash functions (e.g., SHA, MD5, etc.) are used for MAC computations. The Record Protocol can operate without an MAC but is generally only used in this mode while another protocol is using the Record Protocol as a transport for negotiating security parameters.

The TLS Record Protocol is used for encapsulation of various higher-level protocols. TLS Version 1.0 is an Internet protocol based on and very similar to SSL version

3.0. The TLS protocol is misnamed because it operates well above the transport layer (Open System Interconnection [OSI] Layer 4). TLS is used extensively to secure client-server connections on the World Wide Web (TLS is a stateful protocol). Although these connections can often be characterized as short-lived and exchanging relatively small amounts of data, TLS is also in use in environments in which connections can be long-lived, and the amount of data exchanged can extend into thousands or millions of octets. For example, TLS is now increasingly used as an alternative VPN connection. Compression services have long been associated with IPSec and Point-to-Point Tunneling Protocol (PPTP) VPN connections, so extending compression services to TLS VPN connections preserves the user experience for any VPN connection. Compression within TLS is one way to help reduce the bandwidth and latency requirements associated with exchanging large amounts of data while preserving the security services provided by TLS [RFC3943].

10.5 Conclusion

This chapter provided an introductory discussion of the mechanisms that can be utilized to protect the integrity of the information while in transit over an IP-based network, both IPv4 and IPv6. The reader is encouraged to explore further the topic of security at large by consulting appropriate documentation.

References

[ISO198401] International Standards Organization, Information Processing Systems — Open Systems Interconnection Reference Model — Part 1: Basic Reference Model, ISO/IEC 7498-1. (Equivalent to ITU-T Recommendation X.200.) Part 2: Security Architecture, ISO/IEC 7499-2; Part 4: Management Framework, ISO/IEC 7498-4.

[RFC2104] H. Krawczyk, M. Bellare, R. Canetti, HMAC: Keyed-Hashing for Message Authentication, RFC 2104, February 1997.

[RFC2246] T. Dierks and C. Allen, The TLS Protocol Version 1.0, RFC 2246, January 1999.

[RFC2401] S. Kent, R. Atkinson, Security Architecture for the Internet Protocol, RFC 2401, November 1998.

[RFC2402] S. Kent, R. Atkinson, IP Authentication Header, RFC 2402, November 1998.

[RFC2403] C. Madison, R. Glenn, The Use of HMAC-MD5-96 within ESP and AH, REF 2403, November 1998.

[RFC2404] C. Madison, R. Glenn, The Use of HMAC-SHA-1-96 within ESP and AH, RFC 2404, November 1998.

[RFC2405] C. Madison, N. Doraswamy, The ESP DES-CBC Cipher Algorithm with Explicit IV, RFC 2405, November 1998.

[RFC2406] S. Kent, R. Atkinson, IP Encapsulating Security Payload (ESP), RFC 2406, November 1998.

[RFC2407] D. Piper, The Internet IP Domain of Interpretation for ISAKMP, RFC 2407, November 1998.

[RFC2408] D. Maughan, M. Schertler, M. Schneider, J. Turner, Internet Security Association and Key Management Protocol (ISAKMP), RFC 2408, November 1998.

[RFC2409] D. Harkins, D. Carrel, The Internet Key Exchange (IKE), RFC 2409, November 1998.

[RFC2764] B. Gleeson, A. Lin, J. Heinanen, G. Armitage, and A. Malis, A Framework for IP Based Virtual Private Networks, RFC 2764, February 2000.

[RFC2828] R. Shirey, Internet Security Glossary, RFC 2828, May 2000.

[RFC3316] J. Arkko, G. Kuijpers, H. Soliman, J. Loughney, and J. Wiljakka, Internet Protocol Version 6 (IPv6) for Some Second and Third Generation Cellular Hosts, RFC 3316, April 2003.

[RFC3884] J. Touch, L. Eggert, and Y. Wang, Use of IPsec Transport Mode for Dynamic Routing, RFC 3884, September 2004.

[RFC3943] R. Friend, Transport Layer Security (TLS) Protocol Compression Using Lempel-Ziv-Stac (LZS), RFC 3943, November 2004.

[RFC4026] L. Andersson and T. Madsen, Provider Provisioned Virtual Private Network (VPN) Terminology, RFC 4026, March 2005.

[ZHA200401] Y. Zhang, A Multilayer IP Security Protocol for TCP Performance Enhancement in Wireless Networks, IEEE Journal on Selected Areas in Communications, Vol. 22, No. 4, May 2004.

Appendix A: Basic IPv6 Terminology

This appendix provides a glossary of IPv6 terms and concepts, loosely based on Reference [IPV200501] (unless otherwise noted).

6over4	An IPv6 transition technology that provides IPv6 unicast and multicast connectivity through an IPv4 infrastructure with multicast support using the IPv4 network as a logical multicast link.
6over4 unicast address	An IPv6 address of the form: 64-bit prefix:0:0:WWXX:YYZZ, where WWXX:YYZZ is the hexadecimal representation of w.x.y.z, a public or private IPv4 address assigned to the 6over4 device interface.
6over4 link-local address	An IPv6 address of the form: FE80::WWXX:YYZZ, where WWXX:YYZZ is the hexadecimal representation of w.x.y.z, a public or private IPv4 address assigned to the 6over4 device interface.
6to4	An IPv6 transition technology that provides unicast connectivity between IPv6 networks and devices through an IPv4 infrastructure. 6to4 uses a public IPv4 address to build a global IPv6 prefix.
6to4 address	A global IPv6 address of the form: 2002:WWXX:YYZZ:SLA ID: interface ID, where WWXX:YYZZ is the hexadecimal representation of w.x.y.z, a public IPv4 address assigned to a 6to4 router's IPv4 interface. The address space 2002::/16 is assigned to 6to4 addresses.
6to4 host	An IPv6 device that is configured with at least one 6to4 address (a global address with a 2002::/16 prefix). 6to4 devices do not require manual configuration, and they create 6to4 addresses by means of standard autoconfiguration mechanisms.
6to4 router	A router that participates in the 6to4 transition technology, providing unicast connectivity between IPv6 networks and devices through an IPv4 infrastructure.

6to4 relay router	An IPv6/IPv4 router that forwards traffic between 6to4 routers and IPv6 Internet devices.
Address	Network layer identifier assigned to an interface or set of interfaces that can be used as source or destination field in IP datagrams. An IP layer identifier for an interface or a set of interfaces.
	The IPv6 128-bit address is divided along 16-bit boundaries. Each 16-bit block is then converted to a four-digit hexadecimal number, separated by colons. The resulting representation is called colon-hexadecimal. This is in contrast to the 32-bit IPv4 address represented in dotted-decimal format, divided along eight-bit boundaries, and then converted to its decimal equivalent, separated by periods [MSD200401].
	The following example shows a 128-bit IPv6 address in binary form:
	`0010000111011010000000000110100110000000000000000000000010111100111011` `0000000101010101000000000011111111111111110001010001001110001011010`
	The following example shows this same address divided along 16-bit boundaries:
	`0010000111011010 0000000011010011 0000000000000000` `0010111100111011000000101010101010 0000000011111111` `1111111000101000 1001110001011010`
	The following example shows each 16-bit block in the address converted to hexadecimal and delimited with colons:
	`21DA:00D3:0000:2F3B:02AA:00FF:FE28:9C5A`
	IPv6 representation can be further simplified by removing the leading zeros within each 16-bit block. However, each block must have at least a single digit. The following example shows the address without the leading zeros:
	`21DA:D3:0:2F3B:2AA:FF:FE28:9C5A`
Address autoconfiguration	The automatic configuration process for IPv6 addresses on an interface; specifically, the process for configuring IP addresses for interfaces in the absence of a stateful address configuration server, such as Dynamic Host Control Protocol version 6 (DHCPv6).
Address maximum valid time	Time period during which a unicast address, obtained by means of stateless autoconfiguration mechanism, is valid.
Address resolution	Procedure used by a node for determining the link layer address of other nodes on a link. In an IPv6 context, the process by which a node resolves a neighboring node's IPv6 address to its link-layer address. In IPv4, the procedure is accomplished via the ARP protocol. In IPv6, the procedure is accomplished via Neighbor Advertisement and Neighbor Solicitation ICMPv6 messages.

Aggregatable global unicast address	Also known as global addresses, these addresses are identified by means of the three-bit format prefix 001 (2000::/3). IPv6 global addresses are equivalent to IPv4 public addresses, and they are routable in the IPv6 Internet.
Anycast address	A unicast address that is assigned to several interfaces and is used for the delivery of IP datagrams to one of the several interfaces. With an appropriate route, datagrams addressed to an anycast address will be delivered to a single interface — the nearest one.
AS	See Autonomous System.
Attempt address	Unicast address where uniqueness is no longer checked.
Automatic IPv6 tunnel	Automatic creation of tunnels, generally through the use of various IPv6 address formats that contain the IPv4 tunnel endpoints.
Autonomous System (AS)	A network domain that belongs to the same administrative authority.
Bump in the API (BIA)	An IPv6 transition mechanism that performs the IPv4-to-IPv6 translation in the end host. The mechanism works at a layer above IP, translating between IPv4 and IPv6 socket API calls.
Bump in the stack (BIS)	An IPv6 transition mechanism that performs the IPv4-to-IPv6 translation in the end host. The mechanism works at the IP level. BIS employs the SIIT algorithm.
Colon hexadecimal notation	The notation used to represent IPv6 addresses. The 128-bit address is divided in eight blocks of 16 bits. Each block is represented as a hexadecimal number and is separated from the next block by means of a colon (:). Inside each block, left zeros placed are removed. An example of an IPv6 unicast address represented in hexadecimal notation is 3FFE:FFFF:2A1D:48C:2AA:3CFF:FE21:81F9.
Compatibility addresses	IPv6 addresses used when IPv6 traffic is sent through an IPv4 infrastructure. Some examples are IPv4 compatible addresses, 6to4 addresses, and ISATAP addresses.
Compressing Zeros	Some IPv6 addresses expressed in colon-hexadecimal contain long sequences of zeros. A contiguous sequence of 16-bit blocks set to 0 in the colon-hexadecimal format can be compressed to :: (known as double-colon). The following shows examples of compressing zeros [MSD200401]: ■ The link-local address of FE80:0:0:0:2AA:FF:FE9A:4CA2 can be compressed to FE80::2AA:FF:FE9A:4CA2. ■ The multicast address of FF02:0:0:0:0:0:0:2 can be compressed to FF02::2. ■ Zero compression can only be used to compress a single contiguous series of 16-bit blocks expressed in colon-hexadecimal notation.

Correspondent node	Refers to a node that is communicating with a node that is using Mobile IP.
Default route	The route with a ::/0 prefix. The default route is the route used to obtain the next destination address when there are no other matching routes.
Default routers list	A list of routers that can be used as a default router. The list is populated based on Router Advertisement messages received that have a nonnull router lifetime.
Destination cache	Table supported by each IPv6 node that maps each destination address (or address range) to the next hop address to which the datagram has to be sent. It also stores the associated path MTU.
Distance Vector Routing Protocol	A routing protocol in which a router periodically informs its neighbors of topology changes. This is in contrast to link state routing protocols, which require a router to inform all the nodes in a network of topology changes.
Domain Name System (DNS)	A hierarchical storage system and its associated protocol to store and retrieve information about names and IP addresses.
Double Colon	Notation used in compressing continuous series of 0 blocks in IPv6 addresses. For example, the FF02:0:0:0:0:0:0:2 multicast address is expressed as FF02::2.
Dual-Stack Architecture	A node architecture in which two complete protocol stack implementations exist, one for IPv4 and one for IPv6, each with its own implementation of the transport layer (TCP and UDP).
Dynamic Host Control Protocol (DHCP)	A configuration protocol that provides IP addresses and other configuration parameters when connect to an IP network.
Encapsulating security payload	An IPv6 extension header that provides data source authentication, data integrity, and confidentiality.
EUI	See Extended unique identifier.
EUI-64 address	A 64-bit link layer address that is used as the basis to generate interface identifiers in IPv6.
Extended unique identifier (EUI)	Link layer address defined by the Institute of Electrical and Electronics Engineers (IEEE).
Extension headers	Headers placed between the IPv6 header and higher-level protocol headers to provide additional functionalities to IPv6.
Flow	A series of IP datagrams exchanged between a source and a destination.

Format prefix	Variable number of high-order bits of an IPv6 address that defines an IPv6 address type.
FQDN	See Fully qualified domain name.
Fragment	A portion of a message sent by a host in an IPv6 datagram. Fragments contain a fragmentation header to allow reassembly at the destination.
Fragmentation	Process in which the source device divides a message into some number of smaller messages, termed fragments.
Fragmentation header	An IPv6 extension header that contains information that allows the receiving node to reassemble fragments into the original message.
Fully qualified domain name (FQDN)	FQDN gives the full location of a resource within the whole DNS name space. When interpreting the FQDN, one starts at the root and then follows the sequence of domain labels from right to left, going top to bottom within the name space tree. An FQDN includes the top-level domain. For example, www.cnn.com is a fully qualified domain name: www is the host, cnn is the second-level domain, and com is the top-level domain. This is in contrast to a partially qualified domain name (PQDN), which does not give the full path to the domain. One can only use a PQDN within the context of a particular parent domain.
Global address	See Aggregatable global unicast address.
Group identifier	Last 112 bits (for predefined multicast addresses) or last 32 bits (for new multicast addresses) of an IPv6 multicast address used to identify a multicast group [RFC 2373].
Higher-level checksum	A checksum based on the IPv6 pseudoheader; used in ICMPv6, TCP, and UDP.
Higher-level protocol	Protocol that uses IPv6 as transport and is carried as a payload in IPv6, such as ICMPv6, TCP, and UDP.
Hop-by-hop option header	An IPv6 extension header that contains options that must be processed by all intermediate routers as well as the final router.
Host	Any node that is not a router.
Hosts file	A local text file used to contain particular name-to-IP address correspondences. For example, in the Windows® XP operating system the file is located in the \Windowst\system32\drivers\etc directory.
Host-to-host tunnel	An IPv6 over IPv4 tunnel in which endpoints are hosts.
Host-to-router tunnel	An IPv6 over IPv4 tunnel in which the tunnel begins at a host and ends at an IPv6/IPv4 router.

ICMPv6	See Internet Control Message Protocol for IPv6.
Interface	A node's attachment to a link. A representation of a physical or logical link of a node to a link. An example of a physical interface is a network interface. An example of a logical interface is a tunnel interface.
Interface identifier	Last 64 bits of a unicast or anycast IPv6 address.
Internet Control Message Protocol for IPv6 (ICMPv6)	Protocol for Internet Control Messages for IPv6. A protocol that provides error messages for the routing and delivery of IPv6 datagrams and information messages for diagnostics, neighbor discovery, multicast receiver discovery, and IPv6 mobility.
Interworking mechanisms for IPv6 and IPv4	Well-known interworking mechanisms include [RFC2893]: ■ Dual stack: A technique for providing complete support for both protocols — IPv4 and IPv6 — in hosts and routers. ■ Configured tunneling of IPv6 over IPv4: Manually configured point-to-point tunnels for encapsulating IPv6 packets within IPv4 headers to carry them over an IPv4 routing infrastructure. ■ Automatic tunneling of IPv6 over IPv4: Mechanisms for automatically tunneling IPv6 packets over IPv4 networks. ■ Translation: Refers to the direct conversion of protocols.
Intrasite Automatic Tunnel Addressing Protocol (ISATAP)	An IPv6 transition technology that provides IPv6 unicast connectivity between devices placed in an IPv4 intranetwork. ISATAP obtains an interface identifier from the IPv4 address (public or private) assigned to the device. This identifier is used for the establishment of automatic tunnels through the IPv4 infrastructure.
IP6.arpa	The DNS domain created for the IPv6 reverse resolution [RFC 3596]. The reverse resolution has the purpose of "reverse mapping" of IPv6 addresses to DNS names.
IPv4-compatible IPv6 address	A 0:0:0:0:0:0:w.x.y.z or ::w.x.y.z address, where w.x.y.z is the decimal representation of a public IPv4 address. For example, ::131:107:89:42 is an IPv4-compatible address. IPv6 transition mechanisms no longer use the IPv4-compatible address scheme.
IPv6 in IPv4	See IPv6-over-IPv4 tunnel.
IPv4-mapped IPv6 address	A 0:0:0:0:0:FFFF:w.x.y.z (or ::FFFF:w.x.y.z) address, where w.x.y.z is the IPv4 address of an IPv4-only node. Mapped IPv4 addresses are used to represent an IPv4-only host.
IPv4 node	A node that implements IPv4; it can send and receive IPv4 packets. It can be an IPv4-only node or a dual IPv4/IPv6 node.

IPv6 node	Node that implements IPv6; it can send and receive IPv6 packets. An IPv6 node can be an IPv6-only node or a dual IPv6/IPv4 node.
IPv6-over-IPv4 tunnel	Encapsulating IPv6 packets into an IPv4 datagram and transporting the datagram over an IPv4 infrastructure. In the IPv4 header, the Protocol field value is 41.
IPv6 prefixes	The initial bits of an IP address. The number of bits is represented via the prefix-length notation. Prefixes for IPv6 routes and subnet identifiers are expressed in the same way as classless interdomain routing (CIDR) notation for IPv4. For example, 21DA:D3::/48 is a route prefix, and 21DA:D3:0:2F3B::/64 is a subnet prefix. IPv4 implementations commonly use a dotted decimal representation of the network prefix known as the subnet mask. A subnet mask is not used in IPv6. Only the prefix-length notation is used [MSD200401].
IPv6 routing table	Set of routes used to determine the next node address and interface when forwarding IPv6 traffic.
IPv6/IPv4 node	A node that has both IPv4 and IPv6 implementations.
ISATAP	See Intrasite Automatic Tunnel Addressing Protocol.
ISATAP address	An IPv6 address of the form: 64-bit prefix:0:5EFE:w.x.y.z, where w.x.y.z is a public or private IPv4 address allocated to an ISATAP device.
ISATAP device	A device to which an ISATAP address is assigned.
ISATAP name	The name "ISATAP" is resolved by computers with a Windows® XP or Windows Server® 2003 operating system to automatically discover the ISATAP router address for initial configuration.
ISATAP router	An IPv6/IPv4 router that answers ISATAP node requests and routes traffic to and from ISATAP nodes.
Jumbo Payload Option	An option in the Hop-by-Hop Options header that shows the size of the jumbogram.
Jumbogram	An IPv6 packet that has a payload greater than 65,535 bytes. Jumbograms are identified with a 0 value in the Payload Length IPv6 header field and include a Jumbo Payload Option in the Hop-by-Hop Options header.
Lifetime In preferred state	Time during which a unicast address, obtained by means of stateless autoconfiguration mechanism, stays in the preferred state. This time it is specified by the preferred lifetime field in a Router. Advertisement message prefix information option.
Link	A communication facility or medium over which nodes can communicate at the link layer, that is, the layer immediately below IPv6. Examples include Ethernet environments (simple or bridged); PPP links; X.25 Packet Switching; Frame Relay; Cell Relay/Asynchronous Transfer Mode (ATM); or IPv4.

Link State Routing Protocol	A routing protocol in which a router informs all the nodes in a network of topology changes. Information exchanged consists of prefixes of networks connected to the router and their associated cost. This is in contrast to distance vector routing protocols, which exchange routing table information but only with neighboring nodes.
Link-layer identifier	A link-layer identifier for an interface. Examples include IEEE 802 addresses for Ethernet or Token Ring network interfaces and E.164 addresses for ISDN links.
Link-local address	An IPv6 address having a link-only scope, indicated by the prefix (FE80::/10), that can be used to reach neighboring nodes attached to the same link. Every interface has a link-local address.
Local address	An IPv6 unicast address that is not reachable on IPv6 Internet. Local addresses include "link-local" and "site-local" addresses.
Local interface	Internal interface that allows a node to send packets to itself.
Loopback address	The IPv6 address — 0:0:0:0:0:0:0:1 or ::1 — assigned to the local interface.
MAC address	A link layer address for local area network technologies such as Ethernet and Token Ring. It is also referred to as a physical address, hardware address, or network adapter address.
Machine (host)	A node unable to send datagrams not created by itself. A machine (host) is both the source and destination of IPv6 traffic and will discard traffic that is not specifically addressed to it.
Maximum transfer unit (MTU)	Maximum transfer unit (MTU) refers to the size (in bytes) of the largest packet that a given layer of a communications protocol can forward. Maximum transfer units are defined at the link layer (frame maximum size) and at the network or Internet layer (maximum IPv6 packet size).
Maximum-level aggregation identifier	(Also known as top-level aggregation identifier, TLA ID.) A 13-bit field inside the global unicast address reserved for large organizations or ISP by the IANA; hence, it identifies the address range that they have delegated. The TLA scheme has been obsoleted by [RFC3587].
Media Access Control	A sublayer of the link layer of local area networks defined by the Institute of Electrical and Electronics Engineers. Its functionalities include the creation of frames and the management of medium sharing and access.
MTU	See Maximum transfer unit.

Multicast address	An address that identifies several interfaces and is used to deliver data from one source to several destinations. That is, an identifier for a set of interfaces typically belonging to different nodes. By means of the multicast routing topology, packets to a multicast address will be delivered to all interfaces identified by that address.
Multicast group	Set of interfaces listening to a specific multicast address.
Multicast IPv4 tunnel	See 6over4.
Name resolution	Procedure to obtain an IP address from a name.
ND	See Neighbor Discovery.
Neighbor Discovery (ND)	A set of messages and ICMPv6 processes that fixes the relations between neighbor nodes. Neighbor Discovery replaces ARP, ICMP routes discovery, and ICMP redirection messages used in IPv4. It also provides inaccessible neighbor detection.
Neighbor Discovery options	Options in a Neighbor Discovery message that show link-layer addresses, information about prefixes, MTU, routes, and configuration information for IPv6 mobility.
Neighbors	Nodes connected to the same link.
Neighbors cache	A cache supported by each IPv6 node that stores the IP address of its neighbors on the link, its corresponding link-layer address, and an indication of its accessibility state. Neighbors cache is equivalent to the ARP cache in IPv4.
Network address translation — protocol translation (NAT-PT)	Process performed by a network device on the boundary of an IPv4 and IPv6 network. NAT-PT uses a pool of IPv4 addresses for dynamic assignment to the IPv6 nodes. NAT-PT also allows the multiplexing of multiple sessions on a single IPv4 address via the "port" field.
Network addresses translator	A device that translates IP addresses and port numbers when forwarding packets between a network with private addresses and the Internet.
Network segment	See Subnetwork.
Next-level aggregation identifier (NLA ID)	A 24-bit field inside the global unicast aggregatable address that allows the creation of several hierarchical levels of addressing to organize addresses and routing to other ISPs, as well as to identify organization sites. The NLA scheme has been obsoleted by [RFC3587].
NLA ID	See Next-level aggregation identifier.

Non-Broadcast Multiple Access (NBMA)	A link-layer technology that supports links with more than two nodes but without allowing the sending of a packet to all nodes on the link (broadcast). Example technologies include X.25 Packet Switching Service, Frame Relay Service, and Cell Relay Service/Asynchronous Transfer Mode (ATM).
Node	A device that implements IP.
Node types	Node types in an IPv6 environment include the following [RFC2893]: ■ IPv4-only node: A host or router that implements only IPv4. An IPv4-only node does not understand IPv6. The installed base of IPv4 hosts and routers existing before the transition to IPv6 begins are IPv4-only nodes. ■ IPv6/IPv4 node: A host or router that implements both IPv4 and IPv6. ■ IPv6-only node: A host or router that implements IPv6 and does not implement IPv4. ■ IPv6 node: Any host or router that implements IPv6. IPv6/IPv4 and IPv6-only nodes are both IPv6 nodes. ■ IPv4 node: Any host or router that implements IPv4. IPv6/IPv4 and IPv4-only nodes are both IPv4 nodes.
Packet	Protocol data unit (PDU) at network layer. In IPv6, a packet that consists of an IPv6 header and an IPv6 payload.
Parameter discovery	Part of the Neighbor Discovery process that allows nodes to learn configuration parameters, including link MTU, and the default hop limit for outgoing packets.
Path determination	Procedure to select the route from the routing table for use in forwarding the datagram.
Path MTU	Maximum IPv6 packet size that can be sent without using fragmentation between a source and a destination over an IPv6 network route. The route MTU equates with the smallest link MTU for all links in such route.
Path MTU discovery	Process relating to the use of ICMPv6 "Too Big" message to discover the path MTU.
Path vector	A routing protocol approach that involves the exchange of hop information sequences showing the path to follow in a route. For example, BGP-4 exchanges sequences of numbers of Autonomous Systems (ASs).
PDU	See Protocol data unit.

Point-to-Point Protocol	Point-to-point network encapsulation method that provides frame delimiters, protocol identification, and integrity services at the bit level.
Prefix	The initial bits of an IP address. The number of bits is represented via the prefix-length notation.
Prefix length	The number of bits in a prefix.
Prefix-length notation	Notation used to represent network prefix length. It uses the "address/prefix length" form, where prefix length indicates the number of bits in the prefix.
Prefixes list	A collection of prefixes typically used when creating match conditions, for example, for firewall filters.
Protocol data unit (PDU)	A unit of information associated with a particular protocol. During transmission, the protocol data unit of the N layer in a protocol suite becomes the payload of the protocol data unit of the N − 1 layer.
Pseudoheader	Provisional header that is built to calculate the needed checksum for higher-layer protocols. IPv6 uses a new pseudoheader format to calculate UDP, TCP, and ICMPv6 checksums.
Pseudoperiodic	Event that is repeated at intervals of various lengths. For example, the routes advertisement sent by an IPv6 router is made at intervals that are calculated between a minimum and a maximum.
Reassembly	Procedure to rebuild the original message that had been subject to fragmentation.
Redirect	Procedure included in the Neighbor Discovery mechanisms to inform a host about the IPv6 address of another neighbor that is more appropriate as a next-hop destination.
Router	Node that can forward datagrams not specifically addressed to it. In an IPv6 network, a router is also used to send advertisements related to its presence and node configuration information.
Router Advertisement	Neighbor Discovery message sent by a router in a pseudoperiodic way or as a Router Solicitation message response. The advertisement includes, at a minimum, a prefix that can be used by the host to calculate its own unicast IPv6 address following the stateless address configuration procedures.
Router Discovery	Neighbor Discovery process that allows a node to discover routers connected to a particular link.
Router's cache	See Destination cache.
Routing loop	Undesirable situation in a network in which traffic is relayed over a closed loop and never reaches its destination. The Time-To-Live field is used to detect such traffic and delete it.

Scope	For IPv6 addresses, the scope is the portion of the network to which the traffic will be propagated.
Scope ID	The scope ID is an identifier for a specific area or scope.
Site-level aggregation identifier (SLA ID)	A 16-bit field in the global unicast address that identifies subnetworks. The SLA ID field is used by an individual organization to create its own local addressing hierarchy and to identify subnets. The SLA ID scheme has been made obsolete by [RFC3587]
Site-local address	Address identified by the 1111 1110 11 (FEC0::/10) prefix. The scope of these addresses is a local site (of an organization). Site-local addresses are not accessible from other sites, and routers should not direct site-local traffic out of a site. Site-local addresses have been deprecated by [RFC3879]
Solicited-node address	IPv6 multicast address used by nodes during the address resolution process. The solicited-node address facilitates efficient querying of network nodes during address resolution. IPv6 uses the Neighbor Solicitation message to perform address resolution. In IPv4, the ARP Request frame is sent to the MAC-level broadcast, disturbing all nodes on the network segment regardless of whether a node is running IPv4. For IPv6, instead of disturbing all IPv6 nodes on the local link by using the local-link scope all-nodes address, the solicited-node multicast address is used as the Neighbor Solicitation message destination. The solicited-node multicast address consists of the prefix FF02::1:FF00:0/104 and the last 24 bits of the IPv6 unicast address that is being resolved.
Stateless IP/ICMP translation (SIIT)	SIIT is an IPv6 transition technique that allows IPv4-only hosts to talk to IPv6-only hosts.
Static routing	Utilization of routes configured manually into a router's routing table.
Static tunneling	Tunneling technique by which addresses are manually configured for the tunnel source and destination endpoints.
Subnet-router anycast address	Anycast address that is allocated to router interfaces. Packets sent to the subnet-router anycast address will be delivered to one router on the subnet.
Subnetwork	One or more links that use the same 64-bit prefix in IPv6.
TLA ID	See Top-level aggregation identifier.
Top-level aggregation identifier (TLA ID)	See Maximum-level aggregation identifier.
Teredo	IPv6 transition technology for use when IPv6/IPv4 hosts are located behind an IPv4 network address translator.

Teredo Client	Software on an IPv6/IPv4 host allowing it to participate in the Teredo transition technology.
Teredo relay	An IPv6 router that can receive traffic from the IPv6 Internet and forward it to a Teredo client.
Teredo server	A node that assists in the provision of IPv6 connectivity to Teredo clients.
Translation	Translation refers to the direct conversion of protocols, such as between IPv4 and IPv6.
Transport relay translator (TRT)	Transport relay translator partitions the IP layer into two terminated IP legs, one IPv4 and one IPv6. Translation then occurs at the higher layers.
Tunnel	In IPv6 transition context, an IPv6-over-IPv4 tunnel.
Tunneling techniques	Tunneling techniques include the following [RFC2893]: ■ IPv6-over-IPv4 tunneling: The technique of encapsulating IPv6 packets within IPv4 so that they can be carried across IPv4 routing infrastructures. ■ Configured tunneling: IPv6-over-IPv4 tunneling by which the IPv4 tunnel endpoint address is determined by configuration information on the encapsulating node. The tunnels can be either unidirectional or bidirectional. Bidirectional configured tunnels behave as virtual point-to-point links. ■ Automatic tunneling: Tunneling by which the IPv4 tunnel endpoint address is automatically determined, generally embedded in the IPv6 address. Examples include IPv6-compatible addresses and IPv6 6to4 addresses.
Unicast address	An address that identifies an IPv6 interface and allows network layer point-to-point communication. It identifies a single interface within the scope of the unicast address type. An identifier for a single interface. A packet sent to a unicast address is delivered to the interface identified by that address. The following list shows the types of IPv6 unicast addresses: ■ Aggregatable global unicast addresses ■ Link-local addresses ■ Site-local addresses ■ Special addresses, including unspecified and loopback addresses. ■ Compatibility addresses, including 6to4 addresses.
Unspecified address	0:0:0:0:0:0:0:0 or :: — Used to show the absence of any address, equivalent to IPv4 address: 0.0.0.0.

References

[IPV200501] IPv6 Portal, http://www.ipv6tf.org/meet/faqs.php.

[MSD200401] Microsoft Corporation, MSDN Library, Internet Protocol, 2004, http://msdn.microsoft.com.

[RFC2373] R. Hinden, S. Deering, IP Version 6 Addressing Architecture, RFC 2373, July 1998.

[RFC2893] R. Gilligan, E. Nordmark, Transition Mechanisms for IPv6 Hosts and Routers, RFC 2893, August 2000.

[RFC3587] R. Hinden, S. Deering, E. Nordmark, IPv6 Global Unicast Address Format, RFC 3587, August 2003.

[RFC3596] S. Thomson, C. Huitema, V. Ksinant, M. Souissi, DNS Extensions to Support IPv6, RFC 3596, October 2003.

Appendix B:
Basic IPv6 Bibliography

Number	Title	Author(s)	Date		Status
RFC 4779	ISP IPv6 Deployment Scenarios in Broadband Access Networks	S. Asadullah, A. Ahmed, C. Popoviciu, P. Savola, J. Palet	January 2007		Informational
RFC 4773	Administration of the IANA Special Purpose IPv6 Address Block	G. Huston	December 2006		Informational
RFC 4727	Experimental Values In IPv4, IPv6, ICMPv4, ICMPv6, UDP, and TCP Headers	B. Fenner	November 2006		Proposed Standard
RFC 4704	The Dynamic Host Configuration Protocol for IPv6 (DHCPv6) Client Fully Qualified Domain Name (FQDN) Option	B. Volz	October 2006		Proposed Standard
RFC 4692	Considerations on the IPv6 Host Density Metric	G. Huston	October 2006		Informational
RFC 4671	RADIUS Accounting Server MIB for IPv6	D. Nelson	August 2006	Obsoletes RFC 2621	Informational
RFC 4670	RADIUS Accounting Client MIB for IPv6	D. Nelson	August 2006	Obsoletes RFC 2620 Errata	Informational
RFC 4669	RADIUS Authentication Server MIB for IPv6	D. Nelson	August 2006	Obsoletes RFC 2619 Errata	Proposed Standard
RFC 4668	RADIUS Authentication Client MIB for IPv6	D. Nelson	August 2006	Obsoletes RFC 2618 Errata	Proposed Standard
RFC 4659	BGP-MPLS IP Virtual Private Network (VPN) Extension for IPv6 VPN	J. De Clercq, D. Ooms, M. Carugi, F. Le Faucheur	September 2006		Proposed Standard
RFC 4649	Dynamic Host Configuration Protocol for IPv6 (DHCPv6) Relay Agent Remote-ID Option	B. Volz	August 2006		Proposed Standard

RFC	Title	Authors	Date	Notes	Status
RFC 4640	Problem Statement for Bootstrapping Mobile IPv6 (MIPv6)	A. Patel, Ed., C. Giaretta, Ed.	September 2006	Errata	Informational
RFC 4620	IPv6 Node Information Queries	M. Crawford, B. Haberman, Ed.	August 2006		Experimental
RFC 4607	Source-Specific Multicast for IP	H. Holbrook, B. Cain	August 2006		Proposed Standard
RFC 4584	Extension to Sockets API for Mobile IPv6	S. Chakrabarti, E. Nordmark	July 2006	Errata	Informational
RFC 4580	Dynamic Host Configuration Protocol for IPv6 (DHCPv6) Relay Agent Subscriber-ID Option	B. Volz	June 2006		Proposed Standard
RFC 4554	Use of VLANs for IPv4-IPv6 Coexistence in Enterprise Networks	T. Chown	June 2006		Informational
RFC 4489	A Method for Generating Link-Scoped IPv6 Multicast Addresses	J.-S. Park, M.-K. Shin, H.-J. Kim	April 2006	Updates RFC 3306	Proposed Standard
RFC 4487	Mobile IPv6 and Firewalls: Problem Statement	F. Le, S. Faccin, B. Patil, H. Tschofenig	May 2006		Informational
RFC 4477	Dynamic Host Configuration Protocol (DHCP): IPv4 and IPv6 Dual-Stack Issues	T. Chown, S. Venaas, C. Strauf	May 2006		Informational
RFC 4472	Operational Considerations and Issues with IPv6 DNS	A. Durand, J. Ihren, P. Savola	April 2006		Informational
RFC 4449	Securing Mobile IPv6 Route Optimization Using a Static Shared Key	C. Perkins	June 2006		Proposed Standard
RFC 4443	Internet Control Message Protocol (ICMPv6) for the Internet Protocol Version 6 (IPv6) Specification	A. Conta, S. Deering, M. Gupta, Ed.	March 2006	Obsoletes RFC 2463, updates RFC 2780 Errata	Draft Standard

Number	Title	Author(s)	Date		Status
RFC 4429	Optimistic Duplicate Address Detection (DAD) for IPv6	N. Moore	April 2006		Proposed Standard
RFC 4392	IP over InfiniBand (IPoIB) Architecture	V. Kashyap	April 2006		Informational
RFC 4380	Teredo: Tunneling IPv6 over UDP through Network Address Translations (NATs)	C. Huitema	February 2006	Errata	Proposed Standard
RFC 4339	IPv6 Host Configuration of DNS Server Information Approaches	J. Jeong, Ed.	February 2006	Errata	Informational
RFC 4338	Transmission of IPv6, IPv4, and Address Resolution Protocol (ARP) Packets over Fiber Channel	C. DeSanti, C. Carlson, R. Nixon	January 2006	Obsoletes RFC 3831, RFC 2625	Proposed Standard
RFC 4330	Simple Network Time Protocol (SNTP) Version 4 for IPv4, IPv6, and OSI	D. Mills	January 2006	Obsoletes RFC 2030, RFC 1769	Informational
RFC 4311	IPv6 Host-to-Router Load Sharing	R. Hinden, D. Thaler	November 2005	Updates RFC 2461	Proposed Standard
RFC 4305	Cryptographic Algorithm Implementation Requirements for Encapsulating Security Payload (ESP) and Authentication Header (AH)	D. Eastlake 3rd	December 2005	Obsoletes RFC 2402, RFC 2406 Errata	Proposed Standard
RFC 4303	IP Encapsulating Security Payload (ESP)	S. Kent	December 2005	Obsoletes RFC 2406 Errata	Proposed Standard
RFC 4302	IP Authentication Header	S. Kent	December 2005	Obsoletes RFC 2402 Errata	Proposed Standard
RFC 4301	Security Architecture for the Internet Protocol	S. Kent, K. Seo	December 2005	Obsoletes RFC 2401 Errata	Proposed Standard

RFC 4295	Mobile IPv6 Management Information Base	G. Keeni, K. Koide, K. Nagami, S. Gundavelli	April 2006	Errata	Proposed Standard
RFC 4294	IPv6 Node Requirements	J. Loughney, Ed.	April 2006	Errata	Informational
RFC 4293	Management Information Base for the Internet Protocol (IP)	S. Routhier, Ed.	April 2006	Obsoletes RFC 2011, RFC 2465, RFC 2466 Errata	Proposed Standard
RFC 4285	Authentication Protocol for Mobile IPv6	A. Patel, K. Leung, M. Khalil, H. Akhtar, K. Chowdhury	January 2006	Errata	Informational
RFC 4283	Mobile Node Identifier Option for Mobile IPv6 (MIPv6)	A. Patel, K. Leung, M. Khalil, H. Akhtar, K. Chowdhury	November 2005	Errata	Proposed Standard
RFC 4260	Mobile IPv6 Fast Handovers for 802.11 Networks	P. McCann	November 2005		Informational
RFC 4242	Information Refresh Time Option for Dynamic Host Configuration Protocol for IPv6 (DHCPv6)	S. Venaas, T. Chown, B. Volz	November 2005		Proposed Standard
RFC 4241	A Model of IPv6/IPv4 Dual Stack Internet Access Service	Y. Shirasaki, S. Miyakawa, T. Yamasaki, A. Takenouchi	December 2005	Errata	Informational
RFC 4219	Things Multihoming in IPv6 (MULTI6) Developers Should Think About	E. Lear	October 2005		Informational
RFC 4218	Threats Relating to IPv6 Multihoming Solutions	E. Nordmark, T. Li	October 2005		Informational
RFC 4215	Analysis on IPv6 Transition in Third-Generation Partnership Project (3GPP) Networks	J. Wiljakka, Ed.	October 2005		Informational

Number	Title	Author(s)	Date		Status
RFC 4213	Basic Transition Mechanisms for IPv6 Hosts and Routers	E. Nordmark, R. Gilligan	October 2005	Obsoletes RFC 2893	Proposed Standard
RFC 4193	Unique Local IPv6 Unicast Addresses	R. Hinden, B. Haberman	October 2005		Proposed Standard
RFC 4192	Procedures for Renumbering an IPv6 Network Without a Flag Day	F. Baker, E. Lear, R. Droms	September 2005	Updates RFC 2072	Informational
RFC 4177	Architectural Approaches to Multihoming for IPv6	G. Huston	September 2005		Informational
RFC 4159 BCP: 109	Deprecation of ip6.int	G. Huston	August 2005		Best Current Practice
RFC 4147	Proposed Changes to the Format of the IANA IPv6 Registry	G. Huston	August 2005	Errata	Informational
RFC 4140	Hierarchical Mobile IPv6 Mobility Management (HMIPv6)	H. Soliman, C. Castelluccia, K. El Malki, L. Bellier	August 2005	Errata	Experimental
RFC 4135	Goals of Detecting Network Attachment in IPv6	J. H. Choi, G. Daley	August 2005		Informational
RFC 4076	Renumbering Requirements for Stateless Dynamic Host Configuration Protocol for IPv6 (DHCPv6)	T. Chown, S. Venaas, A. Vijayabhaskar	May 2005		Informational
RFC 4074	Common Misbehavior Against DNS Queries for IPv6 Addresses	Y. Morishita, T. Jinmei	May 2005		Informational
RFC 4068	Fast Handovers for Mobile IPv6	R. Koodli, Ed.	July 2005		Experimental
RFC 4057	IPv6 Enterprise Network Scenarios	J. Bound, Ed.	June 2005	Errata	Informational

RFC	Title	Authors	Date	Notes	Status
RFC 4038	Application Aspects of IPv6 Transition	M.-K. Shin, Ed., Y.-G. Hong, J. Hagino, P. Savola, E. M. Castro	March 2005		Informational
RFC 4029	Scenarios and Analysis for Introducing IPv6 Into ISP Networks	M. Lind, V. Ksinant, S. Park, A. Baudot, P. Savola	March 2005	Errata	Informational
RFC 4007	IPv6 Scoped Address Architecture	S. Deering, B. Haberman, T. Jinmei, E. Nordmark, B. Zill	March 2005		Proposed Standard
RFC 3974	SMTP Operational Experience in Mixed IPv4/v6 Environments	M. Nakamura, J. Hagino	January 2005		Informational
RFC 3963	Network Mobility (NEMO) Basic Support Protocol	V. Devarapalli, R. Wakikawa, A. Petrescu, P. Thubert	January 2005		Proposed Standard
RFC 3956	Embedding the Rendezvous Point (RP) Address in an IPv6 Multicast Address	P. Savola, B. Haberman	November 2004	Updates RFC 3306	Proposed Standard
RFC 3919	Remote Network Monitoring (RMON) Protocol Identifiers for IPv6 and Multi Protocol Label Switching (MPLS)	E. Stephan, J. Palet	October 2004		Informational
RFC 3904	Evaluation of IPv6 Transition Mechanisms for Unmanaged Networks	C. Huitema, R. Austein, S. Satapati, R. van der Pol	September 2004		Informational
RFC 3901 BCP: 91	DNS IPv6 Transport Operational Guidelines	A. Durand, J. Ihren	September 2004		Best Current Practice
RFC 3898	Network Information Service (NIS) Configuration Options for Dynamic Host Configuration Protocol for IPv6 (DHCPv6)	V. Kalusivalingam	October 2004		Proposed Standard
RFC 3879	Deprecating Site Local Addresses	C. Huitema, B. Carpenter	September 2004		Proposed Standard

Number	Title	Author(s)	Date		Status
RFC 3849	IPv6 Address Prefix Reserved for Documentation	G. Huston, A. Lord, P. Smith	July 2004		Informational
RFC 3831	Transmission of IPv6 Packets over Fiber Channel	C. DeSanti	July 2004	Obsoleted by RFC 4338	Proposed Standard
RFC 3810	Multicast Listener Discovery Version 2 (MLDv2) for IPv6	R. Vida, Ed., L. Costa, Ed.	June 2004	Updates RFC 2710, updated by RFC 4604	Proposed Standard
RFC 3776	Using IPsec to Protect Mobile IPv6 Signaling Between Mobile Nodes and Home Agents	J. Arkko, V. Devarapalli, F. Dupont	June 2004		Proposed Standard
RFC 3775	Mobility Support in IPv6	D. Johnson, C. Perkins, J. Arkko	June 2004		Proposed Standard
RFC 3769	Requirements for IPv6 Prefix Delegation	S. Miyakawa, R. Droms	June 2004		Informational
RFC 3756	IPv6 Neighbor Discovery (ND) Trust Models and Threats	P. Nikander, Ed., J. Kempf, E. Nordmark	May 2004		Informational
RFC 3750	Unmanaged Networks IPv6 Transition Scenarios	C. Huitema, R. Austein, S. Satapati, R. van der Pol	April 2004		Informational
RFC 3736	Stateless Dynamic Host Configuration Protocol (DHCP) Service for IPv6	R. Droms	April 2004		Proposed Standard
RFC 3701	6bone (IPv6 Testing Address Allocation) Phaseout	R. Fink, R. Hinden	March 2004	Obsoletes RFC 2471	Informational
RFC 3697	IPv6 Flow Label Specification	J. Rajahalme, A. Conta, B. Carpenter, S. Deering	March 2004		Proposed Standard

RFC	Title	Author	Date	Notes	Status
RFC 3646	DNS Configuration Options for Dynamic Host Configuration Protocol for IPv6 (DHCPv6)	R. Droms, Ed.	December 2003		Proposed Standard
RFC 3633	IPv6 Prefix Options for Dynamic Host Configuration Protocol (DHCP) Version 6	O. Troan, R. Droms	December 2003	Errata	Proposed Standard
RFC 3595	Textual Conventions for IPv6 Flow Label	B. Wijnen	September 2003		Proposed Standard
RFC 3590	Source Address Selection for the Multicast Listener Discovery (MLD) Protocol	B. Haberman	September 2003	Updates RFC 2710	Proposed Standard
RFC 3587	IPv6 Global Unicast Address Format	R. Hinden, S. Deering, E. Nordmark	August 2003	Obsoletes RFC 2374	Informational
RFC 3582	Goals for IPv6 Site-Multihoming Architectures	J. Abley, B. Black, V. Gill	August 2003		Informational
RFC 3574	Transition Scenarios for 3GPP Networks	J. Soininen, Ed.	August 2003		Informational
RFC 3572	Internet Protocol Version 6 over MAPOS (Multiple Access Protocol over SONET/SDH)	T. Ogura, M. Maruyama, T. Yoshida	July 2003		Informational
RFC 3542	Advanced Sockets Application Program Interface (API) for IPv6	W. Stevens, M. Thomas, E. Nordmark, T. Jinmei	May 2003	Obsoletes RFC 2292 Errata	Informational
RFC 3531	A Flexible Method for Managing the Assignment of Bits of an IPv6 Address Block	M. Blanchet	April 2003		Informational
RFC 4291	Internet Protocol Version 6 (IPv6) Addressing Architecture	R. Hinden, S. Deering	February 2006	Obsoletes RFC 2373 and RFC 3513	Proposed Standard

Number	Title	Author(s)	Date		Status
RFC 3493	Basic Socket Interface Extensions for IPv6	R. Gilligan, S. Thomson, J. Bound, J. McCann, W. Stevens	February 2003	Obsoletes RFC 2553	Informational
RFC 3484	Default Address Selection for Internet Protocol Version 6 (IPv6)	R. Draves	February 2003		Proposed Standard
RFC 3364	Tradeoffs in Domain Name System (DNS) Support for Internet Protocol Version 6 (IPv6)	R. Austein	August 2002	Updates RFC 2673, RFC 2874	Informational
RFC 3363	Representing Internet Protocol Version 6 (IPv6) Addresses in the Domain Name System (DNS)	R. Bush, A. Durand, B. Fink, O. Gudmundsson, T. Hain	August 2002	Updates RFC 2673, RFC 2874	Informational
RFC 3316	Internet Protocol Version 6 (IPv6) for Some Second- and Third-Generation Cellular Hosts	J. Arkko, G. Kuijpers, H. Soliman, J. Loughney, J. Wiljakka	April 2003		Informational
RFC 3315	Dynamic Host Configuration Protocol for IPv6 (DHCPv6)	R. Droms, Ed., J. Bound, B. Volz, T. Lemon, C. Perkins, M. Carney	July 2003	Updated by RFC 4361 Errata	Proposed Standard
RFC 3314	Recommendations for IPv6 in Third-Generation Partnership Project (3GPP) Standards	M. Wasserman, Ed.	September 2002	Errata	Informational
RFC 3307	Allocation Guidelines for IPv6 Multicast Addresses	B. Haberman	August 2002		Proposed Standard
RFC 3306	Unicast-Prefix-Based IPv6 Multicast Addresses	B. Haberman, D. Thaler	August 2002	Updated by RFC 3956, RFC 4489	Proposed Standard

	Title	Authors	Date	Obsoleted/Updates	Status
RFC 3266	Support for IPv6 in Session Description Protocol (SDP)	S. Olson, G. Camarillo, A. B. Roach	June 2002	Obsoleted by RFC 4566, updates RFC 2327 Errata	Proposed Standard
RFC 3226	DNSSEC and IPv6 A6 Aware Server/Resolver Message Size Requirements	O. Gudmundsson	December 2001	Updates RFC 2535, RFC 2874, updated by RFC 4033, RFC 4034, RFC 4035	Proposed Standard
RFC 3178	IPv6 Multihoming Support at Site Exit Routers	J. Hagino, H. Snyder	October 2001		Informational
RFC 3177	IAB/IESG Recommendations on IPv6 Address Allocations to Sites	IAB, IESG	September 2001		Informational
RFC 3175	Aggregation of RSVP for IPv4 and IPv6 Reservations	F. Baker, C. Iturralde, F. Le Faucheur, B. Davie	September 2001		Proposed Standard
RFC 3162	RADIUS and IPv6	B. Aboba, G. Zorn, D. Mitton	August 2001		Proposed Standard
RFC 3146	Transmission of IPv6 Packets over IEEE 1394 Networks	K. Fujisawa, A. Onoe	October 2001		Proposed Standard
RFC 3142	An IPv6-to-IPv4 Transport Relay Translator	J. Hagino, K. Yamamoto	June 2001		Informational
RFC 3122	Extensions to IPv6 Neighbor Discovery for Inverse Discovery Specification	A. Conta	June 2001		Proposed Standard
RFC 3111	Service Location Protocol Modifications for IPv6	E. Guttman	May 2001		Proposed Standard
RFC 3089	A SOCKS-based IPv6/IPv4 Gateway Mechanism	H. Kitamura	April 2001		Informational

Number	Title	Author(s)	Date		Status
RFC 3056	Connection of IPv6 Domains via IPv4 Clouds	B. Carpenter, K. Moore	February 2001	Errata	Proposed Standard
RFC 3053	IPv6 Tunnel Broker	A. Durand, P. Fasano, I. Guardini, D. Lento	January 2001		Informational
RFC 3041	Privacy Extensions for Stateless Address Autoconfiguration in IPv6	T. Narten, R. Draves	January 2001		Proposed Standard
RFC 3019	IP Version 6 Management Information Base for The Multicast Listener Discovery Protocol	B. Haberman, R. Worzella	January 2001		Proposed Standard
RFC 2928	Initial IPv6 Sub-TLA ID Assignments	R. Hinden, S. Deering, R. Fink, T. Hain	September 2000		Informational
RFC 2921	6BONE pTLA and pNLA Formats (pTLA)	B. Fink	September 2000		Informational
RFC 2894	Router Renumbering for IPv6	M. Crawford	August 2000		Proposed Standard
RFC 2893	Transition Mechanisms for IPv6 Hosts and Routers	R. Gilligan, E. Nordmark	August 2000	Obsoletes RFC 1933, obsoleted by RFC 4213	Proposed Standard
RFC 2874	DNS Extensions to Support IPv6 Address Aggregation and Renumbering	M. Crawford, C. Huitema	July 2000	Updates RFC 1886, updated by RFC 3152, RFC 3226, RFC 3363, RFC 3364	Experimental (published as Proposed Standard)
RFC 2767	Dual Stack Hosts Using the Bump-In-the-Stack Technique (BIS)	K. Tsuchiya, H. Higuchi, Y. Atarashi	February 2000		Informational
RFC 2766	Network Address Translation — Protocol Translation (NAT-PT)	G. Tsirtsis, P. Srisuresh	February 2000	Updated by RFC 3152	Proposed Standard

RFC 2765	Stateless IP/ICMP Translation Algorithm (SIIT)	E. Nordmark	February 2000		Proposed Standard
RFC 2740	OSPF for IPv6	R. Coltun, D. Ferguson, J. Moy	December 1999	Errata	Proposed Standard
RFC 2732	Format for Literal IPv6 Addresses in URLs	R. Hinden, B. Carpenter, L. Masinter	December 1999	Obsoleted by RFC 3986, updates RFC 2396	Proposed Standard
RFC 2711	IPv6 Router Alert Option	C. Partridge, A. Jackson	October 1999		Proposed Standard
RFC 2710	Multicast Listener Discovery (MLD) for IPv6	S. Deering, W. Fenner, B. Haberman	October 1999	Updated by RFC 3590, RFC 3810	Proposed Standard
RFC 2675	IPv6 Jumbograms	D. Borman, S. Deering, R. Hinden	August 1999	Obsoletes RFC 2147	Proposed Standard
RFC 2590	Transmission of IPv6 Packets over Frame Relay Networks Specification	A. Conta, A. Malis, M. Mueller	May 1999		Proposed Standard
RFC 2553	Basic Socket Interface Extensions for IPv6	R. Gilligan, S. Thomson, J. Bound, W. Stevens	March 1999	Obsoletes RFC 2133, obsoleted by RFC 3493, updated by RFC 3152	Informational
RFC 2546	6Bone Routing Practice	A. Durand, B. Buclin	March 1999	Obsoleted by RFC 2772	Informational
RFC 2545	Use of BGP-4 Multiprotocol Extensions for IPv6 Inter-Domain Routing	P. Marques, F. Dupont	March 1999		Proposed Standard
RFC 2529	Transmission of IPv6 over IPv4 Domains Without Explicit Tunnels	B. Carpenter, C. Jung	March 1999		Proposed Standard

Number	Title	Author(s)	Date		Status
RFC 2526	Reserved IPv6 Subnet Anycast Addresses	D. Johnson, S. Deering	March 1999		Proposed Standard
RFC 2497	Transmission of IPv6 Packets over ARCnet Networks	I. Souvatzis	January 1999		Proposed Standard
RFC 2492	IPv6 over ATM Networks	G. Armitage, P. Schulter, M. Jork	January 1999		Proposed Standard
RFC 2491	IPv6 over Non-Broadcast Multiple Access (NBMA) Networks	G. Armitage, P. Schulter, M. Jork, G. Harter	January 1999		Proposed Standard
RFC 2474	Definition of the Differentiated Services Field (DS Field) in the IPv4 and IPv6 Headers	K. Nichols, S. Blake, F. Baker, D. Black	December 1998	Obsoletes RFC 1455, RFC 1349, updated by RFC 3168, RFC 3260	Proposed Standard
RFC 2473	Generic Packet Tunneling in IPv6 Specification	A. Conta, S. Deering	December 1998		Proposed Standard
RFC 2472	IP Version 6 over PPP	D. Haskin, E. Allen	December 1998	Obsoletes RFC 2023	Proposed Standard
RFC 2471	IPv6 Testing Address Allocation	R. Hinden, R. Fink, J. Postel (deceased)	December 1998	Obsoletes RFC 1897, obsoleted by RFC 3701	Historic (published as Experimental)
RFC 2470	Transmission of IPv6 Packets over Token Ring Networks	M. Crawford, T. Narten, S. Thomas	December 1998		Proposed Standard
RFC 2467	Transmission of IPv6 Packets over FDDI Networks	M. Crawford	December 1998	Obsoletes RFC 2019	Proposed Standard
RFC 2466	Management Information Base for IP Version 6: ICMPv6 Group	D. Haskin, S. Onishi	December 1998	Obsoleted by RFC 4293	Proposed Standard

RFC	Title	Authors	Date	Obsoletes/Obsoleted	Status
RFC 2465	Management Information Base for IP Version 6: Textual Conventions and General Group	D. Haskin, S. Onishi	December 1998	Obsoleted by RFC 4293	Proposed Standard
RFC 2464	Transmission of IPv6 Packets over Ethernet Networks	M. Crawford	December 1998	Obsoletes RFC 1972 Errata	Proposed Standard
RFC 2463	Internet Control Message Protocol (ICMPv6) for the Internet Protocol Version 6 (IPv6) Specification	A. Conta, S. Deering	December 1998	Obsoletes RFC 1885, obsoleted by RFC 4443	Draft Standard
RFC 2462	IPv6 Stateless Address Autoconfiguration	S. Thomson, T. Narten	December 1998	Obsoletes RFC 1971 Errata	Draft Standard
RFC 2461	Neighbor Discovery for IP Version 6 (IPv6)	T. Narten, E. Nordmark, W. Simpson	December 1998	Obsoletes RFC 1970, updated by RFC 4311	Draft Standard
RFC 2460	Internet Protocol, Version 6 (IPv6) Specification	S. Deering, R. Hinden	December 1998	Obsoletes RFC 1883	Draft Standard
RFC 2454	IP Version 6 Management Information Base for the User Datagram Protocol	M. Daniele	December 1998	Obsoleted by RFC 4113	Historic (published as Proposed Standard)
RFC 2452	IP Version 6 Management Information Base for the Transmission Control Protocol	M. Daniele	December 1998	Obsoleted by RFC 4022	Proposed Standard
RFC 2450	Proposed TLA and NLA Assignment Rule	R. Hinden	December 1998		Informational
RFC 2428	FTP Extensions for IPv6 and NATs	M. Allman, S. Ostermann, C. Metz	September 1998		Proposed Standard

Number	Title	Author(s)	Date		Status
RFC 2406	IP Encapsulating Security Payload (ESP)	S. Kent, R. Atkinson	November 1998	Obsoletes RFC 1827, obsoleted by RFC 4303, RFC 4305	Proposed Standard
RFC 2402	IP Authentication Header	S. Kent, R. Atkinson	November 1998	Obsoletes RFC 1826, obsoleted by RFC 4302, RFC 4305	Proposed Standard
RFC 2401	Security Architecture for the Internet Protocol	S. Kent, R. Atkinson	November 1998	Obsoletes RFC 1825, obsoleted by RFC 4301, updated by RFC 3168	Proposed Standard
RFC 2375	IPv6 Multicast Address Assignments	R. Hinden, S. Deering	July 1998		Informational
RFC 2374	An IPv6 Aggregatable Global Unicast Address Format	R. Hinden, M. O'Dell, S. Deering	July 1998	Obsoletes RFC 2073, obsoleted by RFC 3587	Historic (published as Proposed Standard)
RFC 2365 BCP0023	Administratively Scoped IP Multicast	D. Meyer	July 1998		Best Current Practice
RFC 2292	Advanced Sockets API for IPv6	W. Stevens, M. Thomas	February 1998	Obsoleted by RFC 3542	Informational
RFC 2185	Routing Aspects of IPv6 Transition	R. Callon, D. Haskin	September 1997		Informational
RFC 2147	TCP and UDP over IPv6 Jumbograms	D. Borman	May 1997	Obsoleted by RFC 2675	Proposed Standard
RFC 2133	Basic Socket Interface Extensions for IPv6	R. Gilligan, S. Thomson, J. Bound, W. Stevens	April 1997	Obsoleted by RFC 2553	Informational

RFC	Title	Authors	Date		Status
RFC 2080	RIPng for IPv6	G. Malkin, R. Minnear	January 1997		Proposed Standard
RFC 2073	An IPv6 Provider-Based Unicast Address Format	Y. Rekhter, P. Lothberg, R. Hinden, S. Deering, J. Postel	January 1997	Obsoleted by RFC 2374	Proposed Standard
RFC 2030	Simple Network Time Protocol (SNTP) Version 4 for IPv4, IPv6, and OSI	D. Mills	October 1996	Obsoletes RFC 1769, obsoleted by RFC 4330 Errata	Informational
RFC 2023	IP Version 6 over PPP	D. Haskin, E. Allen	October 1996	Obsoleted by RFC 2472	Proposed Standard
RFC 2019	Transmission of IPv6 Packets over FDDI	M. Crawford	October 1996	Obsoleted by RFC 2467	Proposed Standard
RFC 1981	Path MTU Discovery for IP Version 6	J. McCann, S. Deering, J. Mogul	August 1996		Draft Standard (published as Proposed Standard)
RFC 1972	A Method for the Transmission of IPv6 Packets over Ethernet Networks	M. Crawford	August 1996	Obsoleted by RFC 2464	Proposed Standard
RFC 1971	IPv6 Stateless Address Autoconfiguration	S. Thomson, T. Narten	August 1996	Obsoleted by RFC 2462	Proposed Standard
RFC 1970	Neighbor Discovery for IP Version 6 (IPv6)	T. Narten, E. Nordmark, W. Simpson	August 1996	Obsoleted by RFC 2461	Proposed Standard
RFC 1933	Transition Mechanisms for IPv6 Hosts and Routers	R. Gilligan, E. Nordmark	April 1996	Obsoleted by RFC 2893	Proposed Standard

Number	Title	Author(s)	Date		Status
RFC 1924	A Compact Representation of IPv6 Addresses	R. Elz	April 1, 1996		Informational
RFC 1897	IPv6 Testing Address Allocation	R. Hinden, J. Postel	January 1996	Obsoleted by RFC 2471	Experimental
RFC 1888	OSI NSAPs and IPv6	J. Bound, B. Carpenter, D. Harrington, J. Houldsworth, A. Lloyd	August 1996	Obsoleted by RFC 4048, updated by RFC 4548	Historic (published as Experimental)
RFC 1887	An Architecture for IPv6 Unicast Address Allocation	Y. Rekhter, T. Li, Eds.	December 1995		Informational
RFC 1886	DNS Extensions to Support IP Version 6	S. Thomson, C. Huitema	December 1995	Obsoleted by RFC 3596, updated by RFC 2874, RFC 3152	Proposed Standard
RFC 1885	Internet Control Message Protocol (ICMPv6) for the Internet Protocol Version 6 (IPv6)	A. Conta, S. Deering	December 1995	Obsoleted by RFC 2463	Proposed Standard
RFC 1884	IP Version 6 Addressing Architecture	R. Hinden, S. Deering, Eds.	December 1995	Obsoleted by RFC 2373	Historic (published as Proposed Standard)
RFC 1883	Internet Protocol, Version 6 (IPv6) Specification	S. Deering, R. Hinden	December 1995	Obsoleted by RFC 2460	Proposed Standard
RFC 1881	IPv6 Address Allocation Management	IAB, IESG	December 1995		Informational

RFC 1827	IP Encapsulating Security Payload (ESP)	R. Atkinson	August 1995	Obsoleted by RFC 2406	Proposed Standard
RFC 1826	IP Authentication Header	R. Atkinson	August 1995	Obsoleted by RFC 2402	Proposed Standard
RFC 1825	Security Architecture for the Internet Protocol	R. Atkinson	August 1995	Obsoleted by RFC 2401	Proposed Standard
RFC 1810	Report on MD5 Performance	J. Touch	June 1995		Informational
RFC 1809	Using the Flow Label Field in IPv6	C. Parttridge	June 1995		Informational

Index